Telecommunications
and Empire

THE HISTORY OF COMMUNICATION

Robert W. McChesney
and John C. Nerone, editors

*A list of books in the series
appears at the end of this book.*

Telecommunications and Empire

JILL HILLS

UNIVERSITY OF ILLINOIS PRESS

Urbana and Chicago

Library of Congress Cataloging-in-Publication Data
Hills, Jill.
Telecommunications and empire / Jill Hills.
p.   cm. — (The history of communication)
Includes bibliographical references and index.
ISBN-13: 978-0-252-03258-5 (cloth : alk. paper)
ISBN-10: 0-252-03258-6 (cloth : alk. paper)
1. Telecommunication—International cooperation—
Political aspects. 2. Telecommunication policy—
United States.
I. Title.
HE7700.H55   2007
384—dc22   2007023284

To Abbey

Contents

Acknowledgments

This book owes a debt to many people. The final manuscript was supported financially by the British Arts and Humanities Research Council. Chapters 3 and 5 came out of research originally supported by the British Economic and Social Research Council and the Nuffield Foundation. The Smithsonian Institution financed archival research at the Museum of American History and at the U.S. National Archives in Maryland. Westminster University financed research at the International Telecommunication Union (ITU) in Geneva.

Some of the ideas in the book have developed from teaching students from almost every country in the world. Others come from ten years of fun collaboration and argument with Bill Wigglesworth and Maria Michalis in the International Institute for Regulators. None bears responsibility for the outcome, but my thanks to all.

In researching the book, Barny Finn was a generous, knowledgeable host at the Museum of American History, where the archivists should be treasured. My thanks to them, to Mary Godwin at Cable & Wireless, and to Heather Heywood at the ITU, who were all helpful beyond the call of duty. Series editor Bob McChesney and Bill Regier at the University of Illinois Press have given me enthusiastic support, and Dan Schiller, together with an anonymous reader, made me rethink the arguments. To all of them, my thanks. Finally, a thank-you to family and friends who have kept me going. The book is dedicated to our youngest member.

Abbreviations

ADAPSO	Association of Data Processing Services Organizations
AID	Agency for International Development
ANC	African National Congress
APEC	Asia-Pacific Economic Cooperation
ARPA	Advanced Research Projects Agency
ARPANET	Advanced Research Projects Agency Network
ASEAN	Association of Southeast Asian Nations
AT&T	American Telephone and Telegraph
AWA	Amalgamated Wireless Australasia Company
CANTAT-1	the first Canadian transatlantic telephone cable
CCIF	International Telephony Consultative Committee
CCIR	International Radio Consultative Committee
CCIT	International Telegraph Consultative Committee
CCITT	International Telegraph and Telephone Consultative Committee
ccTLD	country code top-level domain
CDMA	Code Division Multiple Access
CEPT	Conference Européen de Poste et Telecommunication
CETS	European Conference on Satellite Communications
CEU	Commission of the European Communities
COMPAC	Pacific submarine cable
Comsat	Communications Satellite Corporation
CPT	Compañia Peruana de Teléfonos
CTC	Companiá de Teléfonos de Chile
DGIV	Directorate General IV
DGXIII	Directorate General XIII
DNS	Domain Name Server

EBRD	European Bank for Reconstruction and Development
ELDO	European Launcher Development Organisation
FCC	Federal Communications Commission
GATS	General Agreement on Trade in Services
GATT	General Agreement on Trade and Tariffs
GII	Global Information Infrastructure
GMPCS	global mobile personal communications satellites
GSM	global system for mobile communications
gTLD	generic top-level domain
gTLD-MOU	generic top-level domain memorandum of understanding
IANA	Internet Assigned Numbers Authority
IATA	International Air Transport Association
IBRD	International Bank for Reconstruction and Development
ICANN	Internet Corporation for Assigned Names and Numbers
ICSC	Interim Communications Satellite Committee
IDA	International Development Association
IFC	International Finance Corporation
IFRB	International Frequency Registration Board
IMF	International Monetary Fund
Intelsat	International Telecommunications Satellite Organization
INTUG	International Telecommunications User Group
ISDN	Integrated Services Digital Networks
ITT	International Telephone and Telegraph
ITTA	Information Technology Association of America
ITU	International Telecommunication Union
ITU-D	Development Sector of the ITU
ITU-R	Radiocommunications Sector of the ITU
ITU-T	Standardization Sector of the ITU
KDD	Kokusai Denshin Denwa
LRIC	Long Run Incremental Costs
MFN	Most Favored Nation
MIGA	World Bank's Multilateral Investment Guarantee Agency
MITI	Ministry of Trade and Industry (Japan)
MOU	memorandum of understanding
MPT	Ministry of Posts and Telecommunications (Japan)
NAFTA	North American Free Trade Agreement
NASA	National Aeronautics and Space Administration
NATO	North Atlantic Treaty Organization
NGO	nongovernmental organization
NSAM	National Security Action Memoranda
NSFNET	National Science Foundation Network
NTT	Nippon Telegraph and Telephone

OECD	Organisation for Economic Co-operation and Development
Oftel, Ofcom	Office of Telecommunications (U.K.)
OPEC	Organization of Petroleum Exporting Countries
OSI	Open Systems Integration
PCWATCC	World Administrative Telegraph and Telephone Conference Preparatory Committee
PTOs	Post and Telecommunications Organizations
RASCOM	Regional Satellite for Africa
RCA	Radio Corporation of America
RPOAs	recognized private operating agencies
SAL	structural adjustment loan
SAS	Satellite Business Systems
SES	European Satellite Corporation
SITA	Société Internationale de Télécommunications Aéronautiques
SWIFT	Society for Worldwide Interbank Financial Telecommunication
TAT	transatlantic telephone cable
TAT-1	Transatlantic [telephone cable] No. 1
TEUREM	Tariff Group for Europe and the Mediterranean Basin
TWX	teletypewriter
UNCTAD	United Nations Conference on Trade and Development
UNDP	United Nations Development Program
UNESCO	United Nations Educational, Scientific and Cultural Organization
USITUA	United States ITU Association
USTR	Office of the United States Trade Representative
VADS	Value Added Data Services
VAN	Value Added Network
VANS	Value Added Network Supplier
VSNL	Videsh Sancar Nigam Ltd.
WARC	World Administrative Radio Conference
WATTC	World Administrative Telegraph and Telephone Conference
WATTC-88	World Administrative Telegraph and Telephone Conference/1988
WIPO	World Intellectual Property Organization
WSIS	World Summit on the Information Society
WTO	World Trade Organization

Telecommunications
and Empire

Introduction

This book tells the story of U.S. attempts to liberalize and dominate international telecommunications from World War II until the World Trade Organization agreement of 1997. It is a story about the regulation of the international network—where it was regulated, by what institution, controlled by what power and to whose benefit. It is about the distribution of costs and benefits of regulation among companies and states, and the shifts in dominance of states, companies, and institutions. It is about the interaction of government, markets, and society at national and international levels, and about power—the ability of one state to fashion the international market in its own image. It stretches across the technology of the international telegraph of the first half of the twentieth century to the Internet of the twenty-first.

The book documents the rise and, in some cases, the fall in the regulatory powers of a number of regional and international multilateral intergovernmental institutions—the International Telecommunication Union (ITU), the British Empire/Commonwealth, the Organisation for Economic Co-operation and Development (OECD), Intelsat, the World Bank, the General Agreement on Trade and Tariffs (GATT)/World Trade Organization (WTO), and the European Community (E.U.).[1] However, the story is not simply about multilateral international institutions and their attempts to regulate international telecommunications. It documents U.S. attempts to control, bypass, or replace them with unilateral actions and bilateral agreements and to create private international regulatory institutions, namely, Communications Satellite Corporation (Comsat) and the Internet Corporation for Assigned Names and Numbers (ICANN), directly under U.S. control. The overall thesis of the book is that first in the 1940s, and then from the 1980s on, the United States as the world's dominant economic and military power attempted to restructure the international market of telecommunications to expand its direct and indirect control over the domestic markets of other governments. At the same time, it protected its domestic market from foreign penetration.

The research questions addressed here follow on from *The Struggle for Control of Global Communication,* which dealt primarily with pre–World War II communications. The present book asks, Where did the current system of international regulation of telecommunications come from—on the basis of what shifts in power and to whose benefit? Why the exclusive focus on telecommunications—on person-to-person communication, not broadcasting or other mass media? The answer lies partly in history. Person-to-person communications via the submarine telegraph was the first internationalized form of electrical communication. Telecommunications provided and still provides the infrastructure for companies to spread globally. Until very recently, except for overseas propaganda channels, broadcasting and its regulation have been primarily nationally based.

A second reason for the focus of this book lies in the economic importance of telecommunications, international and domestic, past and present. As those arguing for the inclusion of telecommunications within the GATT negotiations of the 1980s put it, telecommunications was not only an industrial sector in its own right, but it was crucial to providing the underlying means of transportation for trade in services such as finance and audiovisual products. Economically, telecommunications has been much more important to the world economy than broadcasting or film.

This book also provides the opportunity to review the regulation of telecommunications over fifty years in order to bring to light the patterns in the shifts of power between states themselves, between states and international institutions, and between states and companies. It allows us to challenge mainstream views on the origins of international regulation. In particular, it is my contention that the concept of "international regime," hegemonic since the 1980s in any discussion of international regulation, is out of date—that the theory was a product of its time.

The book utilizes primary sources—some British, some American, some from companies, and some from multilateral institutions—and also secondary sources, but it concentrates on aspects that have been only sparingly covered by others.[2] For that reason, the well-documented "New Information Order" debate and the withdrawal of the United States and then Britain from the United Nations Educational, Scientific and Cultural Organization

(UNESCO) is only mentioned. Karl Nordenstreng (1984), Cees Hamelink (1994), and Herb Schiller (1976) lived that debate; I felt that I could contribute little to it. I have gratefully used other secondary literature. Mildred Feldman's (1975) focus on U.S. companies' impact on international telecommunications regulation was ahead of its time. George Codding (1972) and William Drake (1988; 1994) have written a great deal on the International Telecommunication Union. But the chapter here on the World Administrative Telegraph and Telephone Conference of 1988 is written from the perspective of a participant observer. Other than projects supported by the World Bank or written by its personnel (such as those edited by Robert Saunders, Jeremy Warford, and Björn Wellenius [1983] or Björn Wellenius [1993]), there is almost nothing on its role within the international telecommunications market—that by Gwen Urey (1995) and my own unpublished (1993) piece being the exceptions.

Discussion of the role of GATT and the WTO in telecommunications is also limited and often normative. So, for instance, Jonathan Aronson and Peter Cowhey's book (1988) was written for the American Enterprise Institute, and American Express paid for it to be distributed to delegates at the WTO (Mahoney 1993, 339n2). William Drake and Eli Noam's (1997) contrary views on the efficacy of the 1997 Basic Telecommunications Agreement, along with Chantal Blouin's more recent (2000) piece, are useful appraisals. Drake and Noam (1997) and Dwayne Winseck (2002), among others, raise the question of the lack of democracy within trade agreements, and Singh (2002) looks at those agreements from a developing-country perspective, but with the exception of Mathew (2003) there is little on how the Basic Telecommunications Agreement came into being. Brian Winston (1998) writes about the creation of Intelsat in terms of technological innovation, but there is little on global mobile personal communications satellites (GMPCS). In general, therefore, the book should bring together and, I hope, present a pattern and argument missing from the existing literature.

Definitions

The book uses the term "internationalization" not "globalization." Now used in a number of different ways, the term "globalization" was coined in the 1990s to describe a world perceived as characterized by expanding, unaccountable, foreign, direct investment of (particularly) U.S. multinational companies and international capital flows that together affected domestic cultures and economies. Globalization has come to have an ideological, normative dimension in which the process of liberalization of markets is projected as an inevitable process. As Edward Cohen (2001, 14) comments, among those who welcome such a scenario, globalization is seen as "a unified process that is accelerating beyond the control of single nation-states." The crisis of the nation-state became a popular subject in the 1990s (see, for instance, Ohmae 1995). In contrast, academic critics of "globalization" as a concept have questioned its empirical basis, and activists have demanded that sovereign states mitigate its impact (see Hirst and Thompson 1999; Dunkley 1997).

Because "globalization" is such a contested term, the book uses instead "internationalization" to denote the increasing interlinkage and interpenetration of the economies of sovereign states. Electronic communications was the first technological innovation to produce this interlinkage (steam, sea, and air transport followed), so that, from the first international telegraph system, governments had to respond within their domestic economy to events outside their borders. Just as early industrialization caused social problems and the disruptions of boom and bust, so the advent of the almost-instant messaging of the first international telegraph system produced the need for governments to create policy to deal with foreign companies inside their borders and domestic companies wishing to exploit the market outside. The submarine cable license introduced by the U.S. government under President Grant in the 1870s was the first such regulatory control on foreign investment.

Throughout the past centuries electronic communications and foreign direct investment by companies have been interdependent. In the nineteenth century the telegraph reduced the risk of overseas investment by producing immediate information. At the beginning of the twentieth century, wireless technology allowed ships to keep in contact with the shore and made doing business internationally more affordable. Later the inexpensive infrastructure of shortwave wireless allowed direct communications between the colonized and the colonizer. Beginning in the 1960s satellites and submarine telephone cables reduced the costs of person-to-person voice communications and popularized the transnational transmission of live images, thereby making "over there" nearer. In the 1970s, digitalization of communications spurred corporate demand for fast, cheap, international access to data exchange and eased the domestic control of overseas subsidiaries. In the 1990s optic fiber and the Internet allowed the transmission of voice, data, and images, providing the means for both instantaneous capital flows and the outsourcing of services from the industrialized countries to cheaper-labor economies. Yet these developments have not been inexorable. They have depended on the decisions and political priorities of sovereign governments. The argument of this book is that the structure and processes of the current international telecommunications market are the product of regulation, domestic and international.

The term "regulation" is used here to describe a system of power and rules that allows governments to control private corporations and the markets in which they operate. Regulation provides the mechanisms for politics to rule business. Through regulation and regulatory agencies, governments can structure markets and determine the behavior of companies. Domestic regulatory agencies work within the framework of government policy that decides the distribution of costs and benefits. Domestic regulation involves tradeoffs between incumbents and competitors, large users and residential customers, urban and rural, rich and poor. International regulation reaches into the nation-state and provides a framework for this distribution of costs and benefits between states, between states and companies, and between companies and citizens. International regulation demands national implementation and forms boundaries to democracy. Yet, traditionally, international regulation respects the rights of governments of sovereign states to order their own policy priorities. When international regulation does not respect those rights, it constitutes a control of empire.

International regulation can also be a tool of governments. It can be allied with national implementation and national industrial policy to allow sovereign states to support "their" companies—those created under their laws and owned by their nationals—and to control the companies "belonging" to other states in the international market. Strangely, given internationalization of capital and growing international ownership of equity, some states—particularly the United States, France, and Spain—still regard some companies as "their" own. The definition of what is a "national" company still appears based on the place of registration. It is this relationship that is at the root of neomercantilism: expansion abroad and protection at home.

Domestic and International Regulatory Links

Regulation interposes political bureaucracy over corporations and the market. In these early years of the twenty-first century, when we think about the regulation of telecommunications, we tend to think of domestic regulatory agencies, such as the Federal Communications Commission (FCC) in the United States or the Office of Telecommunications (originally Oftel, now Ofcom) in Britain. The FCC, the earliest sector regulator, was created in 1934 at the behest of President Roosevelt in a rushed piece of legislation and was part of the New Deal's concern with the regulation of large corporations in the "public interest." Its aim was affordable telecommunications for all. The FCC, under the control of Congress, was expected to respond to domestic and societal concerns, but by the 1970s, along with other regulatory agencies, it stood accused of failing those it was supposed to benefit.

By the 1970s, after the American Telephone and Telegraph Company (AT&T) had been a virtually unregulated private telephone monopoly for almost a century, U.S. literature on regulation reflected unease. Academics were concerned whether regulation of "natural monopolies" was beneficial to the public interest.[3] They questioned whether regulation reduced prices to consumers, or whether it benefited regulators and the regulated industry. Implicitly, this literature acknowledged that regulation was a mechanism for government to distribute costs and benefits. It accepted the political roots of regulation (see Stigler 1975). Then, in the 1970s and early 1980s, regulatory definitions of markets became political constructs created to benefit new stakeholders seizing the opportunity of new technologies.

During the 1980s, when President Reagan and Prime Minister Thatcher introduced the concepts of small government, privatization of public assets, and the free flow of markets, a number of economists, such as Stephen Littlechild (1983), argued that competition would supplant regulation. In the telecommunications market, AT&T was broken up in a negotiated end to antitrust action and the seven regional Bell Operating Companies created from its local network. Competition was introduced into long distance service. Similarly, in Britain, Mercury PLC was created to compete with British Telecom (later BT)—primarily to service large financial users in London and to compete over long distances.

But the new long distance entrants were weak. It soon became apparent that the old public telecommunications operating monopolies (PTOs) would remain dominant and require considerable ongoing regulation (Benzoni and Svider 1994). The implication of "asymmetric regulation," as it was called, was that the beneficiaries should be the new competitors. Competition became a goal, not a mechanism to a further end.

In Britain the privatization of BT and the creation of Oftel in 1982 separated out operation from regulation. Oftel became a model for other regulatory agencies both in Britain and abroad. Compared to state ownership—the traditional mechanism in Europe for the control of telecommunications networks—the new regulatory agencies, in a surprising development, increased state control. Whereas state ownership had controlled primarily investment and retail tariffs, regulatory authorities used license conditions, price caps (including efficiency gains to the company), and market entry as policy levers.

As Marcellus Snow (1986, 9) notes, "For telecommunications, other dimensions of regulation [were] also important, such as terms and conditions of service (including the obligation of universal service) and requirements for implicit or explicit cross-subsidisation using tariff structures." In other words, telecommunications regulators had a social role to fulfill in their distribution of benefits.

But in Britain the regulatory agency was at first close to the regulated company. Commenting on British experience, Moran (2003, 107) suggests that when the regulatory framework was hurriedly put together, there was a tension between those wishing to see the old British "club system" of governing business maintained and those arguing for a system of rules that would minimize regulatory discretion. However, Whitehall rejected the U.S. system of democratic accountability or perpetual judicial challenge of regulatory decisions. Regulation was intended to take the sector out of domestic politics, particularly out of representative democracy. As Mosco (1990, 51) had pointed out earlier, "Regulation offers representation within a private market structure" to a limited number of participants. But, within a short time and faced with declining quality in the network, Oftel found that without public acceptance, it had no legitimacy. It became more open, both gathering information from the industry and distributing information to the general public (Wigglesworth 1997). It became a political actor with a social remit.

Although regulators might acknowledge their political role, during the 1980s and 1990s the emphasis of much of the academic writing on regulation became economic. The impact of the pervasiveness of the neoliberal "Washington Consensus" promulgated by Wall Street, the World Bank, and others was to obfuscate the political question of why regulate—a question that acknowledged that regulation was a policy mechanism. Instead, regulation came to be more about economic rules and the "neutrality" of the regulatory agency rather than about regulators as political actors (Hills 1991).

Privatization was part of the Reagan and G. H. W. Bush administrations' worldview. Under constant U.S. pressure, the World Bank preached privatization of state-owned assets to poor countries needing debt relief, while at the same time decrying the role of the state. Such an enforced policy benefited Western banks, investors, and operators, and it often transferred control straight back to the previous colonizer's state-owned companies. The World Bank only fully acknowledged that developing countries needed state institutions in 1997—in the same year the WTO Basic Telecommunications Agreement was signed (Hills 1998a). Regulation—with its circumscribed concept of democracy—then became acceptable as a World Bank tool to aid market development. However, the World Bank's telecommunications personnel had been promoting sector regulation through the ITU and their own publications since the 1980s. And it was their promulgation of the benefits of regulation of telecommunications to new market entrants and foreign investors on which the 1997 WTO agreement eventually built.

From the 1940s, U.S. international record companies and Cable & Wireless Ltd. had traditionally worked an end-to-end system for public telegraph networks, collecting traffic overseas and delivering it at home. Under what was called the "special arrangements" clause of the ITU Convention, these privately owned bilateral networks were exempted from international regulation. Coupled with the right to invest and locate in foreign countries, this company-operated end-to-end telegraph network model was originally based on bypassing all international and national regulation.

The introduction of expensive coaxial cable in the 1950s had several effects. It made the old end-to-end transatlantic telegraph lines operated by U.S. international record companies redundant. It also created the additional international capacity for national operators to lease circuits for the private use of large users. Yet at the same time, because national telecommunications monopoly operators (PTOs) shared the costs of the international infrastructure, each route was—for charging purposes—conceptualized as ending halfway. In turn, the concept of "half-circuits" strengthened the basis of state-to-state networks and incumbent PTOs created stiff rules for the use of international-leased "half circuits" so as to prevent private competition.

From the 1980s, as large users gained economic power within the liberalized networks of the United States, Britain, and Japan and were allowed to create their own national data networks, they sought to create private international networks using leased lines. With the growth of foreign investment and reliance on electronic communication, multinational companies wanted to expand their own international, customized company networks. Private networks reduced costs and allowed closer control of overseas subsidiaries. Although in the 1940s it had been the Roosevelt government whose vision comprised a world of sovereign states, the circumstances of the U.S. economy in the 1980s lent favor to the potential economic gains of a borderless international network over which U.S. companies could export their services. But there were a number of impediments. International intergovernmental organizations, international regulations, and national laws stood in the way.

By the early 1980s the United States had begun attempts to liberalize the international market to match up with the domestic market restructuring that had begun in the 1950s. President Reagan unilaterally liberalized Intelsat's state-to-state regulatory system in 1984. That liberalization was followed in 1985 by the start of GATT/WTO negotiations on liberalization of trade in services, with the U.S. intention of replacing Intelsat and the ITU (both controlled by European PTOs) as international regulators. During those on-going negotiations a broad range of U.S. multinational interests attacked the ITU regulations controlling the usage of leased lines. They wanted to impose liberalization not only on the international market, but also on domestic markets worldwide. They attempted to replace the discretionary regulation of sovereign states by the passage of rules promulgated and enforced through international regulation. Such supranational regulation reduced the domestic regulator to an administrator of U.S. policy in what this book terms "Empire rules" regulation.

In some bilateral negotiations—such as those on the North American Free Trade Agreement (NAFTA) with Canada and Mexico—where the United States wielded overriding economic power, they were successful. NAFTA structured Canada's and Mexico's markets to match that of the United States and reduced their discretionary domestic regulation over certain types of new entrant companies to virtually nil. Companies could use private networks for whatever they wanted. And under NAFTA's provisions, corporations were able to sue sovereign states if their corporate plans were interrupted by public policy goals. At the behest of the United States a similar multilateral investment agreement giving corporations rights over sovereign states was drawn up in secrecy within the OECD but failed after public exposure in 1996 (see Barlow and Clarke 1998).

The original U.S. plan for telecommunications services in the WTO was for network liberalization and "Empire rules"-based international regulation, enforced on national governments by a central dispute-resolution mechanism. But in intergovernmental negotiations based on consensus, the United States was unable to push this provision through. Instead, toward the end of the WTO negotiating process the FCC instigated a Regulatory Reference Paper for incorporation in the WTO Basic Telecommunications Agreement. The intention was to ease the process of foreign investment by agreement on basic regulatory principles. But these principles were not "Empire rules" for local administration.

By giving discretion to regulators, the national implementation of the 1997 agreement raised the potential for individual countries to act in favor of their own companies—to bring industrial policy into national regulation or to act in favor of their citizens rather than foreign multinationals. Ironically, under Congressional pressure the FCC was the first to act in this way.

Rather than a globe dominated by U.S. "Empire rules"-based end-to-end networks, the WTO produced a strengthening of the state-to-state network with some domestic liberalization under national control.

Models of International/National Regulation

The very terms "international" and "national" suggest that borders are determinate of policy. They suggest that one thing happens inside and another outside those borders. But the basis of the current study is that there is no such easy division in the telecommunications field. The international telecommunications market is made up of networks, and networks can and do breach borders. Domestic networks can expand into the international, and the international can expand into the domestic. The differences are first in conceptualization and only then in physical characteristics. Regulation—or the lack of it—structures networks. In order to explain the shifts in power within the political economy of international telecommunications over the last sixty years of the twentieth century, we use five models of networks, each benefiting a set of economic and political interests (see Figure 1).

In the first model, from the 1940s to the 1980s, state ownership of telecommunications operators went hand in hand with political independence to create a model of national monopoly and strong state borders. Based on its historical provenance, I call it the "European state model." It benefited the national post and telecommunications administrations (which acted as both operators and regulators), their employees, and national manufacturing industry—what Eli Noam termed the "postal-industrial complex" (Noam 1989). Under U.S. pressure, Intelsat's regulation of international satellite communications was also based on this model—the beneficiaries being the "U.S. satellite-industrial complex."

Second, the "end-to-end model" conceptualizes the network as without regard for national borders. This was the model that inspired the first cross-Channel and transatlantic submarine cables in the 1860s when entrepreneurs expanded domestic public networks to include foreign networks (Hills 2002, 25–28). It involves foreign investment. Because there is no interconnection with another company, such end-to-end transmission allows the greatest profit to network operators. This is the model that International Telephone and Telegraph (ITT) adopted in the 1930s as it linked up national operators it owned in Spain and Latin America with the United States over its own infrastructure. Companies collected traffic abroad and dispatched it directly to the United States using their own networks at each end and their own

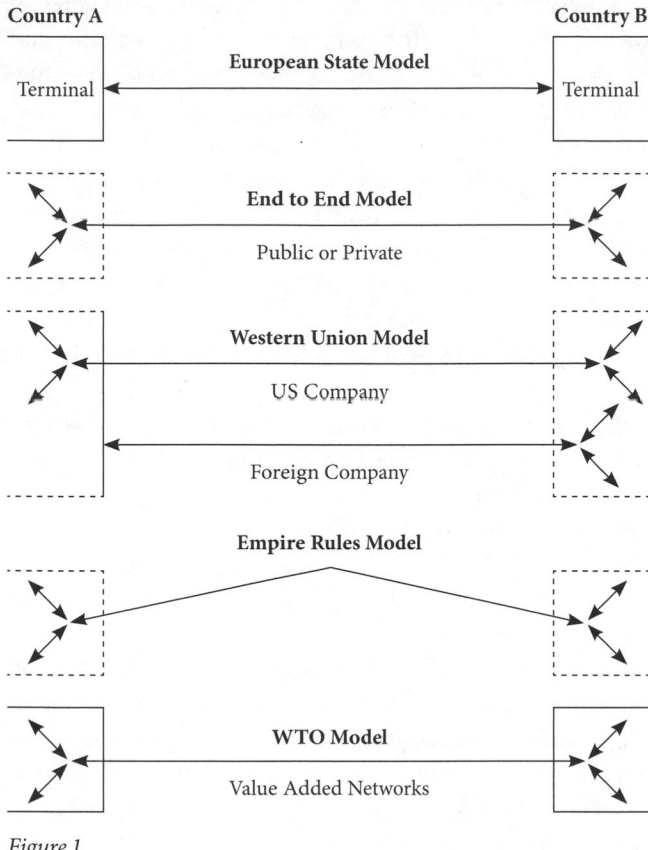

Figure 1

international infrastructure. Foreign governments had little, if any, say over company operations. More recently, the end-to-end model has been associated with the public networks planned first by Comsat and then by global mobile personal communications satellites (GMPCS).

A further variant of the end-to-end model I call the "Western Union model," wherein "international" networks owned by foreign companies are stopped at the borders, while those owned by domestic companies penetrate foreign borders. The name comes from the pre–World War II reality of transatlantic telegraph networks. In 1911 a merged Western Union/AT&T refused to interconnect the British Anglo-American Company's transatlantic lines to the U.S. domestic network and forced the lease of the British company's cables to the U.S. company (Hills 2002, 138–39). At the same time, the U.S.

government pressured the French to allow Western Union and Commercial Cable the same right of establishment as they had in Britain. The Western Union model marries protectionism of domestic markets with free trade rhetoric and can overlap with the "Empire rules" model. It is an asymmetric end-to-end model—the model for neomercantilists. Beneficiaries are those companies protected and promoted by the protectionist state.

Another form of the end-to-end model relates to private networks between multinational companies. The "Empire rules" model represents the outcome of the desire by multinational users to "future proof" their investments against action by sovereign states. But these private networks also serve potential international competitors.[4] Once they collect third-party traffic and interconnect into the public network at each end (termed "international simple resale"), they compete with the state-to-state public networks. Multinational users, beneficiaries of the "Empire rules" model, are those companies seeking to enter the international transmission market and those seeking to sell data-processing equipment overseas.

Then there is the "WTO model," a variant of the state-to-state model, wherein competition may be allowed within the domestic market, but the borders of the state are strengthened. The General Agreement on Trade in Services (GATS) of 1994 accepted that there were a number of possible modes for the delivery of a service that required market access: services could be exported like goods across borders, but they could also require local investment and establishment. However, under GATS (which eventually included the 1997 Agreement on Basic Telecommunications) member states could decide the extent of their commitment and the timing of that commitment. Hence, although GATS instituted competition, it did so under the control of the member state. We characterize the WTO model as similar to that of the European state model, but with some domestic liberalization and increased state control through domestic regulation. Beneficiaries are national regulators and smaller states.

Networks are dependent on interconnection between each other. Each model above suggests different points of interconnection between national and international networks. The European state model suggests interconnection at a specified point designated as the "international gateway," where the national border denotes the end of the international network. The end-to-end model has either no points of interconnection or interconnection within the domestic network at each end. Under the "Empire rules" model, that interconnection is compulsory. The Western Union model suggests asymmetric interconnection—the blocking of access at one end and interconnection within the domestic system at the other. Depending on the undertaking of the specific state, the WTO model allows interconnection for certain services to domestic networks within the state and specifies the terms of those interconnections, supposedly outlawing the inherent discrimination against foreigners in the "Western Union" model.

Standardization affects interconnection. Part of the ITU's function from the nineteenth century forward was to create standards that allowed interconnection between networks for the widest interoperability. These standards allowed interpretation at the national level. Telecommunications operators and governments in the major industrialized countries then used national standards to act as nontariff barriers to foreign penetration of equipment. But digitalization in the 1970s, the overlap with computing technology, and the need for global markets to offset the costs of research and development brought manufacturers into the ITU standardization process. While the de jure standardization of the ITU and other international bodies could foster a global market, proprietary standards (those developed by companies) could be used in an exclusionary manner. Hence, in the 1980s IBM's proprietary System Network Architecture (SNA) standard was in direct competition with the de jure E.U.-inspired Open Systems Integration (OSI) standard.

End-to-end networks favor proprietary standards. State-to-state networks require internationally accepted standards for interconnection. To impose international standards on end-to-end networks is to prevent their being exclusionary and is contrary to the private interests of those who own or run them. Such imposition of ITU de jure standards has been part of the battleground of international regulation between telecommunications and data-processing companies, and more recently between the Internet (data communications) and telecommunications.

Just as points of interconnection and standardization vary, so do the methods of payment for the delivery of traffic. Each method supports different network models. In the nineteenth century, foreign international telegraph companies serving the United States had exclusive interconnection agreements with Western Union or the Postal Telegraph Company, the two U.S. domestic companies, in which they arranged the division of retail tariffs based on relative capital investment or distance. In contrast, the ITU's European state telegraph system used specified tariff rates for the terminal and transit phases of the system that allowed the interconnection on standardized terms of a wide number of networks. Its standardization of tariffs depended on an assessment of the relative costs of transmission and delivery.

What was called the "accounting rate" regime of state-to-state interconnection came into being partly as a result of the pre–World War II campaign of AT&T to challenge the British Empire arrangements that favored Cable & Wireless Ltd. It and other U.S. radio companies negotiated direct routing arrangements with radio telephone and telegraph carriers that bypassed the British system. At first the sender kept two thirds of the revenue. But then in 1938 AT&T reached an agreement with the Australian carrier AWA (Amalgamated Wireless Australasia Company) that the "rates shall be the same in both directions in so far as practicable" and established that "the charges derived . . . shall be divided equally between the parties" (Ergas, Paterson, Geissler 1989, 5). In 1943 the FCC adopted this provision as a precondition for licensing international service to the United States as well as imposing one settlement rate for all the record companies—thereby preventing what was termed "whipsawing" (the playing off of one company against another) by foreign enterprise.

This monopoly-to-monopoly model became institutionalized after the first transatlantic telephone cable was laid in 1956. With fixed exchange rates against the dollar it was possible to use retail tariffs (what the customer paid) based on time and distance as the basis of the system until 1973, when exchange rates became unstable. The retail tariff was then delinked from a wholesale tariff paid by the originating administration for the delivery of its traffic. Partly because the new coaxial cables were often jointly owned by the telecommunications administrations at either end, the network interconnection point was conceptualized as midway along the international circuit. If one operator sent more than 50 percent of the traffic, it owed the other for delivery of the excess. This ITU system, set out in both its Regulations and its Recommendations, supported the state-to-state configuration of the international network.

But the "accounting rate" system also provided a floor price for international communications—state telecommunication administrations could not profitably charge less than the "accounting rate." So, when multinationals began to install their own private networks, either through leased lines or their own infrastructure, they could (by evading international regulation) undercut that floor price. If they then took third-party traffic from other large users and interconnected into the public network at each end, they could provide international "resale" networks. Such "resale" networks would then undercut the "accounting rate" and force it down, cheapening international communications for the industrialized user. Hence the U.S. government, on behalf of its multinationals, set out to gain worldwide "Empire rules" that would ensure leased lines at flat rate prices (not charged according to minutes of traffic), ensure interconnection to the domestic network, allow the use of proprietary standards, allow international "resale," and evade the "accounting rates" systems of payment.

Finally, the way in which "costs" are defined is relevant to the definition of and beneficiaries of networks. At first, under the state-to-state system, how much it "cost" for a national operator to terminate an international call was calculated on what were called "historic" costs—how much had been paid to lay its network using a long depreciation period—and how much users would pay. Traditionally, U.S. companies determined retail tariffs in this demand-based way, and AT&T's "costs" in relation to submarine cables were arrived at on this basis. Then, when competition entered long distance domestic networks, cost allocation changed. "Cost-based" tariffs became fashionable (Oettinger 1988).

This model conceptualized networks as long distance and local. The rubric went that each user must pay the capital "costs" of the local network, allowing new competitors to pay only for usage "costs." The model piled capital "costs" onto access charges to the local network and onto residential users, while it favored new long distance entrants and business. The model was promulgated by the FCC but never fully implemented in the United States and, after initial implementation in the United Kingdom, became politically unacceptable. Instead, the introduction of less drastic "cost-oriented" tariffs required a "rebalancing" of monthly rental, local call, and long distance tariffs—again, to favor long distance (see Kelly 1999). However in the 1990s, with introduction of competition into the international network, AT&T pressed for a "cost-based" division between international and national networks, so that international users of the public network would not pay toward the capital cost of developing countries' networks. International users of the public network would have benefited at the expense of the developing world. In contrast, international users had to pay a "universal connectivity charge" to interconnect with AT&T.

"Costs" of interconnection have become a political issue as new companies have entered former monopoly markets. Originally left to company-to-company agreements, the United States has pressed for international regulation and a particular formula of what is termed Long Run Incremental Costs (LRIC)—pricing according to marginal costs (Oftel 2002). The FCC has disagreed with other countries on such arcane features as the "costs" of capital to go into the calculation of LRIC (see, for instance, Motohiro and Thierer 2003). Low interconnection rates favor "resale" as a mechanism for introducing competition—pushed by the FCC—whereas higher interconnection rates favor a strategy of creating infrastructure (favored by some U.S. states).

The Internationalization of Domestic Regulation

Much of this book is about the conceptualization of alternative market structures, about whose interests regulatory definitions benefit. Just as a division within an end-to-end network between "international" and "national" is necessary if the networks are to be differentially regulated, so the end to end domestic network has to be similarly divided if new entrants are to be allowed. The regulator can divide the network by geography into intrastate and interstate (or into international and domestic) local and long distance; or by "costs"—thick and thin lines, urban and rural; or by equipment—consumer and capital; or by message type—voice, telegraph, fax, data; or by technology—fixed wire, wireless, cellular mobile, satellite, Internet. The conceptualization in one major market tends to be followed elsewhere.

Peter Cowhey argued in 1989 that "countries seek international rules for markets to reinforce their domestic regulatory arrangements" (200). The more powerful the state, the more these domestic provisions are likely to be exported into international agreements, thereby making domestic and international regulation compatible. So, for instance, the 1875 St. Petersburg Treaty of the International Telegraph Union replicated the provision of the British Telegraph Act of 1869 that had excluded from public ownership private company-to-company networks. The ensuing "special arrangements" for bilateral networks excluded them from the regulation of the ITU and became important to the United States from the 1940s on.

In similar fashion the rather haphazard way in which the U.S. domestic market was restructured from the 1950s to allow entry to AT&T's end-to-end domestic monopoly was then replicated in the United Kingdom. So the British government first liberalized the equipment market, then the long distance market, and then, following the FCC's Computer II decision in 1980, divided up the remaining network according to the "value added" to transmission. The intention was similar to that in the United States, where IBM pressed to enter the transmission network and was awarded a division of the network into "basic" (roughly voice) and "enhanced" (roughly data) services. Neither was technologically divisible, and such regulatory divisions were an empirical nonsense. Nevertheless, the services were regulated differently, with "enhanced" or "value added" regulated much less than basic or voice services and not required to contribute to the costs of universal service. Eventually, service suppliers providing "enhanced" services were allowed to interconnect both ends into the public network and take third-party traffic, thereby creating an alternative "resale" network for large users.

Because "enhanced services" or "value-added network" suppliers (VANS) evaded U.S. domestic regulation they obviously wanted to ensure that international regulation did not give the right to other countries to regulate them. The economic rationale for these new services relied not only on favorable regulation but also on the leasing of capacity from network operators at bulk prices. As a result, the availability and cost of "leased lines" and how they should be charged to users featured as a worldwide regulatory issue from the 1970s on.

In view of the ITU's regulation of the international network on the basis of the "European State" model, the new "enhanced" services were first allocated to the existing category within the ITU regulations of "special arrangements" that allowed them to escape the "accounting rates." Cowhey (1989, 210–11) terms this regulatory space within the ITU regulations as the "grey market of private negotiations"—equivalent to a black market in foreign exchange. France and Spain, through the ITU's 1988 World Administrative Telegraph and Telephone Conference (WATTC), unsuccessfully attempted to turn the clock back. Then followed the GATS agreement of 1994, in which governments made commitments to allow these "enhanced services." In effect, the regulatory divisions created in response to the domestic politics of the United States in 1980 were institutionalized first in Japan and the United Kingdom, then in Intelsat and the ITU, and finally in the WTO.

The Endgame of the WTO

The WTO 1997 Agreement on Basic Telecommunications and its Regulatory Reference Paper took fifteen years to negotiate, during which time the structure of the international telecommunications market and those of industrialized economies had greatly changed. This book will argue that during that period the international model of regulation pursued by the United States was asymmetric regulation based on a combination of the "Western Union" model of the prewar period and an "Empire rules" model. This combination demanded alteration in the regulation governing international networks and the control foreign governments had over investment to allow U.S.-controlled end-to-end networks. But through domestic regulation of a bygone era that protected the U.S. market from overseas competition—the Buy American Act of 1933 and the provisions of the 1910 Wireless Act—the model prevented or slowed penetration of foreign companies into the United States. Congress, the FCC, and the Office of the United States Trade Representative (USTR) provided the nexus for political lobbying by economic interests for this combination model.

In the years preceding the entry of the United States into World War II, it was the international record companies and AT&T that formed the strongest lobby. In the 1950s AT&T became preeminent, but from the 1960s, as domestic liberalization began, multinational user companies gathered strength. The small group of companies such as Citibank, American Express, and IBM, which pushed in the 1980s through the USTR for the introduction of telecommunications into GATT, intended that U.S. companies should gain rights of investment overruling national sovereignty and that the agreement should provide worldwide enforcement of rules to ensure end-to-end private networks. But they had lost economic and political clout by the 1990s.

Pressure for the liberalization of Intelsat's maturing market in the 1980s came from U.S. satellite manufacturers who were seeking exports. In addition, Panamsat Corporation and Global Personal Communications Satellite operators that included investors, such as Motorola, demanded end-to-end public networks and protection from Intelsat and European operators. Later, in the potential that privatized networks gave for equipment exports, the Europeans viewed the Clinton administration's concept of the liberalized Global Information Infrastructure as a response to the U.S. electronics manufacturing industry. Later still, AT&T, which had been failing in the postdivestiture chaos of the 1980s, once more gained political momentum in its traditional alliance with the FCC. These changes serve to demonstrate how the strength of U.S. corporate lobbies, each allied with different state agencies, rose and declined and often competed with each other. Nevertheless, under the U.S. trade legislation of 1974 and 1988, the relationship between the USTR, Congress, and corporations became ever closer as legislation provided the means to apply political pressure to foreign governments.

Hence, the public interpretation put on the WTO agreement by the FCC suggesting that the rest of the world had accepted the U.S. 1996 Telecommunications Act as the "gold standard" had implications in U.S. trade-related complaints.[5] The USTR went on to produce numerous complaints against other countries for not implementing the provisions of the 1996 Act. But the WTO did not apply to any service that was not specified in a country's commitment for liberalization on a certain date. And, even then, the manner of liberalization and ensuing market structure could be determined by the national regulator.

In fact, the U.S.-inspired Regulatory Reference Paper incorporated by countries into their GATS commitments produced a set of principles that aided the transparency of domestic regulation. It did not give foreign entrants the universal right of establishment, insist on liberalized markets, or prevent state ownership of the operator. It did not necessarily apply to the Internet. Even in terms of equal treatment for foreign and national companies, it did not prevent regulators from issuing demands on domestic companies that indirectly affected foreign companies. And it did not prevent the FCC from making such a unilateral declaration as in its 1996 benchmark order on "accounting rates," discussed in chapters 4 and 6.

The contention of this book, written in 2007, is that the U.S. strategy of enforcing the liberalization of worldwide national telecommunications markets so that U.S. companies might operate end-to-end control has only been partially successful. Its defeat has come from opposing international alliances and conflicts of interest within the U.S. political system. Similarly, its attempts to restructure, bypass, and create international institutions directly under its control have had only limited success. Nevertheless, the book argues that the U.S. "Empire project" is still ongoing.

Theoretical Framework

The mainstream theoretical approach to international regulation since the 1980s has been encompassed in the concept of "international regime." Regime theory arose in the early 1980s in a period of relative decline in U.S. economic power. In what became known as the "hegemonic stability" or "lost hegemony" thesis, political scientists questioned whether global free trade to the benefit of U.S. interests could be sustained without a hegemon (the United States) to control other states. Such questions occupied U.S. theorists until the 1990 Gulf War and the fall of the Soviet Union, when the United States became the overt economic and military superpower.

Krasner's answer was to put forward "international regimes" as the means to gain consensus on outcomes that would favor U.S. interests. He defined regimes as "sets of implicit or explicit principles, norms, rules and decision-making procedures around which actors' expectations converge in a given area of international relations" (1983b, 3). The concept provided an appealing hierarchy; Drake (1994, 184–85), for instance, uses it to discuss the legal instruments of the International Telecommunication Union—the Convention, Regulations and Recommendations—equating each with one level of Krasner's typology.

Much has subsequently been written on international regimes in numerous issue areas, such as international post, international shipping, the environment, outer space, and others (Zacher and Sutton 1997; Vogler 2000). To realist international relation theorists, "regimes" came about as the product of power relations between nation-states and resulted from diplomacy and game-theory negotiating ploys. To neorealist theorists such as Cowhey (1990), regimes are the outcome of negotiation within international institutions that reflected the interplay of domestic interests on the foreign policy of states. Others have pointed to the influence of shared knowledge and values on regime formation (Young and Osherenko 1995, 225)

In the early 1990s, at a time when the "Washington Consensus" was prevalent, the concept of an "epistemic community" became popular. In such a community, people held ideas in common that shaped international agendas and prevented opposing views from being aired (Haas 1992). In turn, this cognitive approach to regimes married with the network theories gaining prevalence in British political science (Rhodes and Marsh 1992). Following on from this concern with ideas and networks "characterized by trust and mutual adjustment," the regime concept has tied in with the late twentieth-century concept of "governance"—also a contested term often used to express a view that government goes beyond formal mechanisms of parliament and voting to include wider groups in society (Rhodes 1996, 653). John Vogler (2000) broadens the concept in this way: he argues that international relations theorists are not interested in international organizations *per se* but in the whole range of informal agreements, networks, and values that constitute the mechanisms for international collaboration in the governance of the issue areas of satellite orbits and radio spectrum. A "regime" is a pattern of governance; hence, the concept of "regime" moved from that of an institution to that of a "pattern" with backward links to the 1970s literature on "interdependence" theory and its ideas of "issue areas" and networks of bureaucracies (Keohane and Nye 1977; Keohane and Milner 1997).

In telecommunications the emphasis has been on using the concept as an explanatory tool, rather than questioning the concept itself. I have also used it in this manner (1998b). In general, the mainstream view of international telecommunications regards international regulation as encapsulated by the international "regime" of the ITU (Cowhey 1990; Lee 1996; Melody 2000). Here the idea is that the shared engineering background and PTO "club" of the participants was an important component in the international agreements reached (Ergas and Pogorel 1994, 27). By implication, that "club" declined, power switched from those participants, a new "club" arose, and beneficiaries are now different.

Overwhelmingly, what emerges from those who actually try to utilize the concept is the view that a "regime," whether institution or network, denotes consensus. But consensus hides the exercise of power. Its very creation benefits one set of interests over another. So, for instance, within the ITU in the 1970s it was possible for the United States to utilize the tradition of "consensus" to overcome the numerical majority of developing countries and for developing countries to use it against the United States in GATT in the 1980s.

If we accept that a "regime" indicates consensus, then there are problems in the concept's explanatory force of our empirical data. Until the late 1940s the so-called "international" regime failed to include the most powerful economy—the United States. Do we ignore what the United States was doing during this period? Or do we acknowledge that it made unilateral declarations on domestic telegraph matters that impinged directly on foreign-owned companies, or that the FCC attempted to establish itself as a global regulator? These actions were not consensual. There may have been a "club" inside the United States, but it belonged only peripherally to one outside.

Similarly, how do we explain U.S. attempts to set up unilateral private corporate international regulatory institutions under U.S. control, such as Comsat in the 1960s for satellites and ICANN in the 1990s for the Internet? How do we explain forum shopping—the bypassing of one international regulatory institution for another: the ITU and OECD for GATT, the United Nations for Comsat and Intelsat, the ITU for ICANN? And how do we explain U.S. bilateral and unilateral actions throughout the century, each attempting to replace international agreements with the unilateral projection of U.S. law or regulation onto the international stage? In *Struggles for Control of Global Communication* there are numerous examples of such behavior. In the present work, examples stretch through the 1940s to the 1980s and 2000. None of these actions was aimed at an international consensus around "norms" or "principles." Rather, they were direct actions by U.S. state agencies to ensure that the interests of U.S. corporations were preeminent.

Nor within the original concept of "international regime"—as a variable between states and outcomes—is there a place for the leadership role of the officials of the international institution itself. Do their actions come under the rubric of "shared norms," or does it make more sense to see their actions as placating their most powerful members? The literature on WTO and World Bank personnel seems to suggest the latter, and I take that view in my discussion of the ITU in chapters 3 and 4 (see also Fatoumata and Kwa 2003).

Finally, where do markets and companies fit into the regime concept? Are markets simply part of the consensus around which the regime coalesces? That possibility positions markets as free-floating entities, whereas my argument is that they are conceptual creations resulting from regulation or the lack of it. As for companies, do they simply act as a lobby pushing governments into international agreements in their favor, taking their (governments') place on international delegations and even writing the international agreement? This book documents how they have done all these things. But both Comsat and ICANN have had international regulatory power devolved on them by the U.S. government—both entering into conflict with other governments in the process.

It would be possible to expand the concept of "international regime" to include every action and idea by individual governments, companies, and international bureaucrats, but it gives us few clues as to how to predict further changes. The argument here follows Susan Strange (1983), who contended that, by focusing on the mechanism of regime formation within international institutions, theorists deflected attention away from regime content and beneficiaries, and regime theory underplayed American structural power. But, in a traditional political economy approach, Strange equates structural power with resources, whereas we see power in the *use* of those resources, not in the resources *per se*. Defined in her way, the concept of structural power has problems, as Steven Lukes (2004), its originator, has acknowledged: it tends to be less than specific as to how the hegemon sets the agenda and uses its resources to determine outcomes.

Instead, in the context of a half century of conflict between the United States and the rest of the world over its determination to prevent sovereign states from exercising that sovereignty over U.S. multinational corporations, it is useful to characterize U.S. power in terms of the fourfold classification of direct and indirect power used by Barnett and Duvall (2005). The present argument starts from the fact that as the United States gained economic power in the twentieth century, and the transatlantic route became the engine of international telecommunications traffic, so it was necessary for other countries to interconnect with the United States for trade and investment flows. It was these factors that gave the U.S. government its structural centrality and direct power in world telecommunications. Direct relational power (power of A to get B to do what A wants) feeds off direct structural power—the power that stems from the use of resources such as control of entry into the U.S. market. But in addition, indirect power can come from positioning within institutions and through the promulgation of ideas—what is sometimes called "soft power." So, for instance, in its 1940s desire to alter international regulation, the FCC was able to exercise not only direct relational power over U.S. companies, but also worldwide indirect power over foreign companies wishing to interconnect with U.S. companies. Later, the FCC had both direct power over Intelsat in its control of contracts in favor of U.S. industry and indirect power through its control of transatlantic submarine cables. It was able to set the possibilities of the future for Intelsat.

Ideas have been important as well. In the 1980s when the United States was the first market to undergo the data processing challenge to the incumbent telecommunications monopoly, the FCC was able to influence the ideas of how to structure domestic networks. Until the European Union, in its Green Paper of 1987, arranged a division between "reserved" services and others, the U.S. domestic regulatory structure had become the "standard." In turn, structuring of the international market to replicate the underlying model of U.S. domestic regulation both reinforced the centrality of the U.S. telecommunications market and expanded the U.S. market overseas to benefit large users and suppliers in the United States.

In the WTO negotiations, again, the FCC had the capacity to shape future possibilities through its institutional centrality and its instigation of the regulatory provisions of the WTO Basic Telecommunications Agreement. But evidence seems to show that indirect power has not been enough. U.S. negotiators supplemented bargaining by direct threats of retaliation, by the implicit threat of the creation of regional agreements such as NAFTA, and by withdrawal in the interests first of AT&T and then of GMPCS operators. My argument is therefore that throughout the fifty years covered here, U.S. agencies used direct relational and structural power and indirect institutional and ideational power to gain control of world telecommunications for U.S. interests—to attempt to create a U.S. Empire.

Overall, this book starts when the British Empire was at its peak and concludes at the end of the twentieth century, when the U.S. Empire was in the making. We use the term "empire" to denote a system of international relations that puts the United States at the center of a web of international lines of communication, public and private, from which it is intended that the U.S. economy and favored companies shall benefit. There are shades of comparison with the British Empire's use of communications, to dominate economically and/or militarily in some cases, to exercise indirect economic control in others, to extract resources from yet others, and to lock out potential communications competitors. Although security and telecommunications policy overlap at various times (for instance AT&T, ITT, Intelsat, and GMPCS operators, such as Iridium, were closely connected to defense) and the dimension of social control through electronic communications has become increasingly important over time, what concerns us here is not a military empire *per se,* but an economic empire that allows the colonizer to place the resources of the colonized at its disposal. The argument follows

Chalmers Johnson (2000) who interprets the period since World War II to be one of an unacknowledged U.S. "Empire project." He contends, "From 1992 to 1997, the United States led an ideological campaign to open up the economies of the world to free trade and the free flow of capital across national borders. Concretely this meant attempting to curb governmental influence, particularly any supervisory role over commerce in all 'free market democracies'" (207).

In arguing that U.S. actions from the 1940s to 1997 in the communications sector were designed to create an empire, there is a danger in characterizing policy as a process that goes uninterrupted from A to B with one articulated goal, rather than a series of sometimes unrelated, haphazard events in which domestic as well as international actors fight each other for less than noble reasons. Hence, in explaining U.S. actions, it is necessary to take into account competition between agencies, both executive and representative, within domestic political systems, and the impact of particular personnel, both at international and national levels. International agreement can legitimize domestic actions. "Policy laundering" is now a term used to denote the means by which governments can gain policy outputs through international agreement and then impose at home, in the name of that international agreement, what they could not achieve through domestic processes (Hosein 2003). As Nye (2004, 161) points out, "Government officials shop among forums as they try to steer issues to arenas more favorable to their preferred outcomes; and they use international organizations as instruments to bring pressure on other governments as well as other departments of their own governments." International agreements can be the mechanism for strengthening the domestic power of one regulatory agency or another, or one set of domestic interests over another. New agency personnel are likely to be most active in this respect.

Nor in my argument concerning a U.S. Empire do I deny that the United States failed sometimes in its use of power. It failed in the FCC attempt to open the submarine cable market in 1980 to international "resale"; in 1988, when it was unable to prevent ITU Member States from retaining the recognized right to operate state-owned networks; and in the 1990s, when Asian governments refused to open their markets to GMPCS operators. Direct and indirect power of the United States could be defeated when the rest of the world banded together.

But the concepts of power, empire, and control allow us to understand why U.S. demands have brought it into conflict with the developing world and into conflict with U.N. institutions based on the concept of sovereignty.

Noam Chomsky, writing just after the first Gulf War, argued that the demise of the Soviet Union replaced an East-West conflict with a North-South conflict. It created the perception in the Global South of a unipolar world in which U.S. control was exercised, not only through the military, but through the U.S.-dominated Bretton Woods international financial institutions (Chomsky 1991). This book documents World Bank policy with its impact of recolonization and, how, having insisted that the ITU become a U.N. institution in the 1940s, much U.S. action since the 1980s has been concerned with containing the influence of developing countries and undermining the ITU's remit.

One can argue, as Krasner did in 1985, that international institutions, particularly those based on one nation–one vote, are themselves a potential source of opposition to the United States. But Peter Cowhey's statement that the intention of the United States in GATT was to divest not only the U.N.-based ITU but also Intelsat of regulatory power suggests that the problem rests not with U.N. institutions but with any institution over which the United States does not have direct control. Particularly since the 1980s, it is the international intergovernmental institution itself that has threatened American corporate capitalism and its control of the South. Indirect control by the United States has not been enough—hence, the directly controlled, private Comsat and ICANN.

This book argues that to a large extent we can explain the rise and fall of the regulatory power of international institutions in terms of their control by the United States. At the end of the twentieth century the ITU had become little more than an agency of the WTO. Where the ITU still held some regulatory power—as over satellite orbits—there were calls by the FCC chairman for privatization (Powell 2004). Also, indirect power over the World Bank has not satisfied successive U.S. administrations. The United States has attempted to constrain the World Bank's actions by (unsuccessful) attempts to alter its charter and by perpetual criticism. The World Bank appears to have lost influence as private financing of investment has weakened its multilateral lending function. Once the enforcer of liberalization and privatization on developing countries, in 1997 the World Bank also came up against the institutional turf of the WTO and lost power to a more directly U.S.-controlled international institution.

But the WTO agreement, partly because of the actions of the United States itself, did not deliver the end-to-end control of networks U.S. multinational companies wanted. Instead, it gave regulators in developing countries additional power. And, with the economic rise of China, India, and Brazil, developing countries began to demand more from the developed. The U.S.

response was traditional—eschewing the multilateral for the unilateral and bilateral and the public regulator for the private under U.S. control. Finally, ending the old order of international regulators, Intelsat privatized itself in 2001 (see Hills 2007). The U.S. government created a private international institution under direct U.S. government control—ICANN—for the U.S.-based data network—the Internet. Under the George W. Bush administration, unilateral action through domestic regulation and bilateral action through free trade agreements have followed.

The overall theoretical framework of this book, therefore, is that, driven by societal pressures, congressional fury, fragmented agencies, and corporations able to work the system, U.S. direct and indirect power has been directed in telecommunications toward breaking down regulatory barriers so as to create a world in its own market image. That world would give U.S. corporations end-to-end networks on terms that would prevent any foreign domestic regulation, give its manufacturing industry equipment exports, and give its service industries global rights of investment over sovereign states. That world would use telecommunications for the global expansion of U.S. economic control. But that "Empire project" has been hampered by the WTO, by the negotiating strategies of developing countries, and by the conflict between domestic agencies and their clients. The year 1997 and the WTO Basic Telecommunications Agreement marked the beginning of a new round of the U.S. "Empire project."

Overview

At the beginning of World War II there were three regulatory blocs governing international telecommunications: the United States, the British Empire, and the ITU. Chapter 1 covers FCC attempts to liberalize the monopoly of the British Empire over British resistance and to extend its control unilaterally to South America and countries in "liberated" Europe. It documents British efforts to defend its commonwealth system, to rid itself of the end-to-end U.S. transatlantic telegraph carriers, and to prevent the ITU becoming an agency for U.S. equipment manufacturers.

Chapter 2 discusses the impact of the first transatlantic telephone cable jointly owned by AT&T and the Canadian and British governments in 1956, and British attempts to reincarnate its global cable hegemony. The U.S. government then preempted those developments by unilateral regulation of satellite communications through the private corporation, Comsat. Although a U.S. public campaign opposed such private operation, only European recalcitrance led to an international operator/regulator, Intelsat. The major beneficiaries of Comsat's and Intelsat's regulation of the world satellite market were U.S. satellite manufacturers. But AT&T's opposition and U.S. domestic regulation in AT&T's favor meant that the FCC became the unseen regulator of the world international telecommunications market.

Chapter 3 analyzes ITU international regulations and the U.S. attempt to restructure those regulations from the European state model toward the end-to-end model. From 1980 on, the United States moved to export its regulation for the benefit of financial institutions and data processing companies, such as IBM. Then at the World Administrative Telegraph and Telephone Conference of 1988, the U.S. government demanded international regulation that would have exported liberalization into national markets. Although it failed, that moment represented the virtual end of the monopoly system of international communications. The European Union gave notice that the Treaty of Rome's espousal of competition overrode the ITU's regulatory structure.

Chapter 4 analyzes changes in the power structure within the ITU. The rise in the number of member states following decolonization, in turn, created demands for equity. These demands were consistently thwarted by U.S. and European delegations until restructuring of the organization reduced the power of the South. From the late 1980s, the ITU came under the influence of the World Bank and preached privatization, liberalization, and regulation on the Bank's behalf. Declining finances also obliged the ITU to give more power to the private sector. In the 1990s, the United States bypassed the ITU in its unsuccessful bid to establish end-to-end networks through GMPCS operators. It then humiliated the ITU first in the FCC's unilateral regulation of "accounting rates" and later in the creation of another private regulator directly under U.S. control—ICANN.

Chapter 5 looks at the World Bank's role in the regulation of international and national telecommunications. Often galvanized by U.S. pressure, the World Bank became a major force in the privatization of state assets in developing countries—privatization that opened markets to foreign investors and promised rollback of the state. However, from the 1980s on, telecommunications personnel pressed for state regulatory agencies on the British model. Inadvertently, they institutionalized a state-led model; in 1997, the bank recognized state regulation.

Chapter 6 discusses the fifteen years of WTO negotiations. The WTO Telecommunications Agreement of 1997 was thought to define international and national regulation of telecommunications so as to underpin an end-to-end network model. But rather than liberalizing national markets for the benefit of U.S. companies and end-to-end operation, the WTO agreement empowered national regulators and developing countries. Even when Mexico was caught out by unnoticed provisions of the WTO Reference Paper, the agreement could not force the domestic creation of "resale" networks so desired by U.S. operators. Instead, it strengthened the regulatory power of weak developing states and centralized power in federal states.

Overall, this book documents how ideas and issues have been transferred from one international institution to another, empowering one and weakening the other; from OECD to GATT (transborder data flows); from GATT to NAFTA and back again (value-added networks); from GATT to the ITU (global mobile personal communications satellites); from the World Bank to the WTO (liberalization); and from WTO to OECD (investment). By 1997 there were still several regulators of international networks—the European Union and the WTO being the most powerful.

The book, then, is a complex mix demonstrating that international regulation comprises not only state actions in negotiations within international institutions, but also bilateral and unilateral initiatives and company-to-company agreements. Communications were the glue that held the British Empire together. This book documents the partially successful attempt over half a century to create a similar U.S. Empire. That attempt has had societal impacts both inside the United States and worldwide. If the book aids debate on this overall "Empire project," then the effort has been worthwhile.

1

Opening Up the British Empire

This chapter tells the story of U.S. attempts to liberalize the British Empire monopoly, to bring the International Telecommunication Union (ITU) under its influence, and to retain the international record companies' transatlantic end-to-end networks. Wartime powers allowed the Federal Communications Commission (FCC) to restructure and control the domestic and international telegraph market and to unilaterally expand its regulation within the U.S. sphere of influence. Then, undermined by congressional attacks, it lost power to the international record companies and the Pentagon.

Although the FCC and International Telephone and Telegraph (ITT) attempted to preserve a "Western Union" model of end-to-end telegraph networks, the first transatlantic telephone cable institutionalized a state-to-state model and strengthened the domestic monopoly of American Telephone and Telegraph (AT&T).

Background

From the declaration of war between Britain and Germany on September 2, 1939, and Pearl Harbor on December 7, 1941, relations between the United States and Britain were uneasy. The U.S. administration suspected that aid to Britain would support the postwar retention of the British Empire. In response, in the Atlantic Charter on August 14, 1941, Prime Minister Churchill and President Roosevelt agreed "to see sovereign rights and self government restored to those who have been forcibly deprived of them." It marked the beginning of the end of the British Empire. The subsequent 1942 Mutual Aid Agreement that formed the basis for Lend-Lease reiterated these objectives, and the United Nations, established in 1945, built on the idea of sovereign rights.

Spurred by the belief that the 1919 Wilson administration had been unprepared for peace, the Roosevelt administration began "drawing blue-prints of the post war world as early as the autumn of 1939" (Gardner 1956, 5). The British civil servant John Maynard Keynes is said to have taken a draft plan for a "new order" to the United States in a 1941 visit (Harrod 1972, 10). In the Bretton Woods discussions of 1944 that led to the International Bank for Reconstruction and Development (World Bank) and International Monetary Fund, Sir Roy Harrod, a confidant of Keynes, recalled "the postwar negotiations as involving a trade-off between American interest in commercial policy and British interest in international monetary arrangements" (cited in Acheson, Chant, and Prachowny 1972, xiii). Fearing a postwar recession, the U.S. government wanted free trade for its manufactured goods, a favorable climate for U.S. foreign investment, and unimpeded access to raw materials.

In the immediate postwar period, the threat of Communism led to the June 1947 Marshall Plan for the reconstruction of Europe. In the three-and-a-half years of the plan, about $13 billion was spent in Europe, much of it returning in purchases of U.S. food and machinery (Galbraith 1994, 159–60). Such anticommunist concern also stretched to the developing countries. In 1949, in his inaugural address, President Truman outlined his Point Four Program under which the United States would give assistance to developing countries. Galbraith comments, "American economic policy would be influenced by a continuing combination of compassion, idealism, and paranoiac fear of Communism" (Galbraith 1994, 162). Churchill first used the term "Iron Curtain" in 1946; in the same year, Hans Morgenthau, previously Roosevelt's Secretary to the Treasury, introduced the model of "realism" and "national interest" into international relations (Morgenthau 1946). The cold war had begun.

International Telecoms Market Structure

The world telecommunications market prior to World War II was characterized by three contrasting international market structures—that of the ITU, that of U.S. companies, and that of the British Empire. The ITU system worked on the basis of the "European state" model, exchanging traffic at the borders. The U.S. system was a mixture of end-to-end working, the "Western Union" model, and state-to-state working. The British Empire system was a monopoly end-to-end system.

The ITU structure, in place since the nineteenth century, was governed by a convention and a series of regulations on technological standardization for each technology—telegraph, radiotelegraph, and radiotelephone. The European telegraph regulations extended to the French colonies in North Africa and those companies registered or receiving a landing license in a member state, but the European telephone regulations extended only to Western Europe. The extra-European telegraph regulations covered Canada, South America, and the Far East, but not the United States.

Three technical committees, loosely organized within the ITU, oversaw standardization—The International Telegraph Consultative Committee (CCIT), the International Radio Consultative Committee (CCIR), and the International Telephony Consultative Committee (CCIF).[1] In order to create a worldwide interconnecting system, the ITU's standardization had to reach down into national provision, and for this reason companies and PTOs from non-ITU members could attend the committees (AT&T attended the prewar CCIF, for example). When interconnecting with European states, the U.S. international record companies had worked under the essential ITU regulations since the 1880s but resisted U.S. membership of the ITU. As a member state, the United States would have had to apply ITU regulations—for instance, implementing standardized classification of telegraph traffic—to its companies. The exception to this distancing by U.S. interests was the radio sector, where U.S. manufacturing and shipping interests needed the ITU for frequency allocations.

From the first U.S. international telegraph cable installation in 1881 the U.S. telegraph companies had worked on an end-to-end basis. Western Union and later the Commercial Cable Company (through its domestic partner the Postal Telegraph Company) had picked up traffic in the United States and delivered it to company offices in Britain, there collecting traffic for the return journey. Originally, the French cable companies delivered traffic along their own U.S. domestic lines, but then Western Union took these over, and from 1910, when the "Western Union" model of transatlantic communications came into being, no foreign company was allowed to distribute traffic in the United States. The end-to-end model was also the norm for prewar ITT. The company had bought domestic operating companies in Rumania, Spain, Peru, Chile, Puerto Rico, Argentina, and Brazil, mainly to gain procurement for its equipment factories, and from 1929 interconnected them to its domestic U.S. subsidiary, Postal Telegraph.

The technology of longwave and shortwave radiotelegraph escaped U.S. domestic regulation until 1927. Radio Corporation of America (RCA), established in 1919, sometimes operated both ends of an international link, but it tended to construct point-to-point operations on the basis of the "European state" model. Whereas U.S. cable companies had been forced in 1920 to accept competition on their routes, only after 1936 was RCA obliged to liberalize its monopoly landing licenses. However, this provision did not apply to AT&T.

AT&T began to transmit overseas telephone calls in 1927; by 1936, as well as having a virtual monopoly on domestic telephone transmission and a share of the domestic telegraph market, the company had opened twenty-seven direct international radiotelephone circuits. During the 1920s AT&T established interconnection with telephone systems in Cuba, Canada, and Mexico without any form of U.S. license. High international tariffs, low traffic volume, and the ITU system all reinforced the company's international voice monopoly.

The British Empire system was created in 1937 under Article 13 of the ITU Madrid Convention that allowed "special arrangements." The "special arrangements" clause originally allowed bilateral arrangements to be made "of every kind in matters of service which do not interest the generality of states."[2] The British Empire system worked on the basis that the whole conglomeration of Dominions and colonies was one international end-to-end bloc served by one monopoly international carrier—Cable & Wireless Ltd. For the U.S. companies this bloc was an impediment to global expansion, and the European war provided the necessary political leverage to open up the British Empire.

The U.S. Domestic Market

In the U.S. domestic market from 1935 on, it became evident that the two domestic telegraph companies, Western Union and Postal Telegraph, had financial problems caused by mounting competition from telephone, airmail, AT&T, and the military. By 1938, Western Union's revenue had dropped by almost 40 percent and the smaller Postal Telegraph's by 20 percent. AT&T's revenue from its leased teletypewriter (TWX) lines for corporate users was equivalent to a quarter of Western Union's total revenue and 18 percent of domestic traffic—2 percent more than Postal Telegraph's share.[3] In addition, in 1939 the Army Administration Communications System and the Naval Communications System handled commercial and government traffic totaling more than 30 percent of civil messages.[4] Under the provisions of the 1860 Post Roads Act, the U.S. government was also able to impose tariffs 40 percent below commercial rates for its telegraph traffic. As a result of these factors, in 1939, to prevent its bankruptcy, Postal Telegraph received $5 million from the Reconstruction Finance Corporation.

International and national were linked. The 1934 Communications Act split international record companies according to the technology used—wireline or radio. Western Union was both a domestic and international wireline company with ten transatlantic submarine cables (five leased from the Anglo-American Telegraph Company). The ITT group of companies included Postal Telegraph, All America Cables and Radio Inc. (serving predominantly South America and the Caribbean), the Commercial Cable Company (operating six transatlantic cables), and American Cable and Radio Corporation with interests in Pacific cables. Of the radio companies, RCA Communications Inc. was the largest, with fifty two international radiotelegraph circuits. Its smaller rival Mackay Radio and Telegraph Co., one of the ITT group, operated ten circuits to Russia and China, among others. French Cable Company, the only foreign-owned cable company with transatlantic cables, ceased operation during the European war.

James Fly, a lawyer, had moved from the Tennessee Valley Authority to become chairman of the FCC in 1939. He was a fervent "New Dealer." Arguing that the downward trend of the domestic telegraph was inevitable, he proposed to Congress the amalgamation of the Postal Telegraph and Western Union, together with a takeover by Western Union of AT&T's TWX service.[5] But Congress hesitated, fearful of labor union opposition and of creating a domestic telegraph "monopoly."

During this debate Chairman Fly argued that the international radio telegraph companies' practice of establishing numerous international gateways spread throughout the country was adversely affecting the domestic carriers. He proposed that international radiotelegraph and cable companies should be merged and confined to the international realm, thereby separating domestic from international networks.[6] Congress remained unconvinced. Meanwhile, profiting from wartime traffic, the international record carriers were making money.

Direct Radio Circuits to the Empire

Until 1939 the FCC had refused to allow competition between radiotelegraph circuits, allowing only intermodal competition with U.S. cable routes. But war in Europe caused radio traffic to drop. To compensate for declining revenue, the radio companies demanded that the U.S. government use its possible entry into the war as a lever to open the British Empire market.

The British Empire market structure had been created in 1928 when the British government forced the merger of the Eastern Telegraph Company and the Marconi Wireless Telegraph Company. For security reasons it had decided to preserve high-cost cable against low-cost shortwave wireless. But the merged company, Cable & Wireless Ltd., could not prosper under the weight of debt it accrued. In 1937 the government restructured that debt, accepting as a quid pro quo from the company a low, uniform "Empire Rate" tariff that was independent of distance. From 1937, Cable & Wireless adopted this preferential flat rate tariff of 1s 3d (fifty-two cents) for the British Empire and 1d (three cents) per word for the press.

In order to appease the Dominion governments who saw Cable & Wireless as an inefficient monopoly, in 1928 the British government also placed the company under an Imperial Telecommunications Advisory Board (renamed the Commonwealth Communications Council in 1942) made up of representatives from Britain and the Dominions. The board had some regulatory control over the company's tariffs in that it could veto rate increases. But it lacked a remit on either "costs" or investment, so that the company, which both funded and housed the board in company headquarters, could easily frustrate it. Consequently, the company had poor relations with the board chair, Sir Campbell Stuart, a Canadian.

The U.S. radio companies first attempted to break into the British Empire when, in 1928, in order to evade high cable tariffs, the Amalgamated Wireless Australasia Company (AWA), jointly owned by the Australian state and the Marconi company, negotiated an agreement with Colonel Sarnoff of RCA for a direct radiotelegraph circuit with Sydney. But at the subsequent 1928 Empire conference, the Australian government offered a guarantee, renewed in 1937, that it would not allow any such direct wireless links.[7] Then relations between the Australian government, which aimed to fully nationalize AWA, and Cable & Wireless deteriorated. Poor relations with both Australia and Campbell Stuart worked against the company.

Even after Pearl Harbor the British refused direct U.S. radio circuits to the Empire. Cable & Wireless argued that the U.S. companies would cherry-pick the most lucrative routes. However, once Congress had passed Lend-Lease in February 1942 and after the intervention of Harry Hopkins (President Roosevelt's personal assistant), the then Imperial Advisory Board decided in favor of the U.S. companies.[8] The FCC began to allow competition between radio companies on the same routes. By the summer of 1943 Mackay and RCA had opened twenty-five new international radiotelegraph circuits, some of which, such as those to Gambia and British Guiana, carried almost no traffic.

The U.S. companies pushed to carry transit traffic to third countries over their new lines and a 1942 Commonwealth meeting, dominated by Australia and Campbell Stuart, decided in their favor. Clement Attlee informed Cable & Wireless of the decision a year later.[9] The company then began a rearguard action to persuade the British Cabinet that to allow transit traffic would "carry with it the transfer of the news centre of the world to New York away from London" and was "part of a plan to break up the communications of the Empire."[10] It argued that whereas the potential loss to the company from the U.S. direct circuits was £300,000 annually, if they were allowed to carry "transit" traffic—for instance, if the India-to-U.S. circuit could also carry U.K.-bound traffic—then the loss would be nearly £1 million annually.

The problem for the British government was whether Cable & Wireless could survive postwar competition. By the end of the war in Europe in 1944 the submarine cable system was in a state of disrepair. With only one repair ship available and both steel and skilled labor scarce, the company had set up radio services to plug gaps in the system, but Commonwealth countries had to share resources on a regional basis.[11]

A major concern to the British government was the extent of the U.S. military radio communications network. As *Fortune* commented in 1944, the U.S. Army Signals Corps appropriations since 1941 amounted to $9 billion and "the army now has the largest wire and radio system the world has ever seen and is likely to end the war with more radio telephone and telegraph equipment than all the private companies put together."[12] The British press had also reported U.S. government radio engineers had "worked out a tentative plan for world radio communications after the war. There would be great trunk lines around the world at about 20 degrees north of the Equator."[13] The British ambassador to Washington urged the Commonwealth governments to adopt a definite telecommunications policy "before the Americans strengthen their own position by effecting a merger of their international telegraph undertakings."[14]

Therefore the question of Cable & Wireless's relations with the Dominions coincided with questions about FCC policy that affected the company's future profitability. AT&T's attempts to establish an end-to-end radio service to Egypt (then part of the British informal empire) sounded warning bells. To the British it seemed that the company was attempting to please the FCC.[15] Countervailing pressures—from the Dominions for autonomy, from the U.S. radio companies for transit traffic, and from the security and defense services for the retention of cables—led to nationalization as a method of saving the company.

In April 1944, the company proposed a scheme to bring all the international communications facilities in the Empire under one entity—an Empire corporation—on which all the Commonwealth governments would be represented. But the Dominions' desire for national autonomy sabotaged such centralization. They proposed an "Anzac" scheme that decentralized ownership. However, because that scheme would have allowed the Dominions and India to establish direct telegraph circuits to foreign countries, a Cabinet committee accepted Cable & Wireless's arguments that it was contrary to the Commonwealth undertakings of 1937.

Eventually the cabinet sent Lord Reith (previously Director General of the BBC), not Campbell Stuart, whom they distrusted, to negotiate a compromise with the Dominions. Despite company opposition, the Commonwealth Telecommunications Conference of July 1945 recommended that "the governments of the United Kingdom, the Dominions and India should acquire the private shareholdings in the overseas telecommunications services of those countries, and the shareholdings of Cable and Wireless in the Dominion and Indian companies; and that those companies should be re-organised on a uniform basis."[16] But the conference supported the company's view that U.S. direct circuits should be allowed only where the volume of traffic justified them and that they should not carry transit traffic. In other words, the Empire system would be defended against the U.S. entrants, but its defense would be based on national control of national assets.

Just before the November 1945 Bermuda talks (see below), the new British Labour government accepted a variant of the Canberra proposals. Under this scheme the Commonwealth Communications Council, now renamed the Commonwealth Telecommunications Board, was to be given wider functions, including the development of a round-the-world cable system.[17] The board was to be "the first global telecommunications management entity" and a potential Commonwealth regulator (Oslund 1977, 152). The company was to retain its cable system together with domestic and external communications in the colonies until their independence (its monopolies in Hong Kong and the West Indies being most profitable). However, the Commonwealth partner governments, including the British, would acquire the company's assets in their countries. Two weeks later the U.S. dropped the atomic bomb and Japan surrendered.

The FCC and a State-to-State Model

Immediately after the Japanese attack on Pearl Harbor, on December 7, 1941, President Roosevelt signed over control of radio communications to the Defense Communications Board (later succeeded by the Board of War Communications). In March 1942 he also gave the board control over wire communications, thereby increasing the power of the chairman, James Fly, who also retained his position as chair of the FCC.[18]

For security reasons, the board cancelled the FCC policy of competitive circuits and reduced the routes by almost half (FCC 1943, 36). In June 1942, ostensibly to facilitate censorship of international telegraph messages, the board also ordered RCA and Mackay Radio to limit entry to three international "gateway" cities—New York, Washington, and San Francisco.[19] These limitations on international gateways remained until the 1980s. The decision separated international from national and benefited the domestic telegraph companies.

In May 1943 the international carriers were then refused permission to negotiate with foreign administrations. Only when the FCC had awarded a license under competitive tender could the carrier start negotiations (FCC 1943, 35). In addition, the FCC made it a condition of international licenses that international settlements be made on a 50/50 basis (FCC 1943, 37 and 40). Hence, the FCC used its position within the Board of War Communications to extend its regulatory powers over the companies.

In the spring of 1942 President Roosevelt supported the introduction of legislation to permit the Western Union/Postal Telegraph merger, but he requested that post offices be sited in telegraph offices and that there should be protection for displaced labor.[20] Chairman Fly once more urged the merger of the international companies but was vociferously opposed by Rear Admiral Stanford C. Hooper, one of the founding spirits of RCA in 1918.[21] For Hooper and the trade unions, amalgamation would benefit ITT, considered a "foreign" company because it was controlled by non-Americans and suspected of collaboration with the Nazis.

Signed by President Roosevelt on March 9, 1943, the eventual act divided international from domestic telegraph. It terminated the end-to-end transatlantic system of Western Union. Although it allowed the merger between Western Union and Postal Telegraph, it required that "any such merger should provide for the divestment of the international operations of any merged company."[22] Western Union had to sell its international business but not to an international carrier—ruling out its acquisition by "foreign" ITT.

Underlying this separation of domestic from international was fear on the part of the international record companies that Western Union, with both a domestic monopoly and international facilities, could discriminate against them. In order to avoid that possibility, the 1943 act brought into being a "just, reasonable, and equitable formula in the public interest."[23] This formula was, in effect, an FCC cartelization of international telegraph traffic. It was to define the market structure of the U.S. transatlantic international telecommunications system for more than four decades and to have repercussions into the 1990s.

The FCC fixed the share of future outgoing international traffic that each company could transmit according to its 1942 share of the U.S. international market. If incoming "unrouted" traffic increased (where customers did not specify a company), then the companies received an equivalent proportion of outgoing traffic—in other words, there was little that companies could do to undermine the cartel.[24] As shown in Table 1, shares of the U.S. international market in terms of revenue hardly altered between 1946 and 1951, the increase in that of All-America Cables being primarily due to an overall increase in South American traffic over which it had a virtual monopoly.

According to the FCC, the general principle behind the formula was that "conflicting obligations for the distribution of traffic should be resolved by dividing the traffic between the interested carriers on an equitable basis."[25] The companies self-regulated the agreement through an International Formula Committee with a permanent secretariat (the International Quota Board), but they perpetually complained about each other's behavior.[26] Because of continued opportunities for discrimination in favor of Western Union's own international facilities, from 1946 the FCC forced the company to conduct its international business through a separate department—Western Union Cables.[27]

Table 1. Proportion of Gross Revenue of U.S. International Telegraph Companies, 1945–50

Company	1946	1947	1948	1949	1950
WUC	19.5	18.6	20.5	20.2	19.1
CCC	11.4	11.6	10.3	9.7	9.0
AACR	22.2	25.0	24.4	23.9	25.4
MRT	12.3	15.2	13.9	16.1	15.7
RCAC	34.0	29.5	30.8	30.1	30.7

Note: WUC: Western Union Telegraph Company (Cable Division); CCC: American Cable & Radio Corp. Subsidiaries–Commercial Cable Company; AACR: All America Cable and Radio Inc.; MRT: Mackay Radio and Telegraph; RCAC: RCA Communications.

Source: Derived from "Table III: Comparison of Gross and Net Operation Revenues. Major Carriers—International Industry." In Telecommunications: A Program for Progress; A Report by the President's Communications Policy Board. March 1951, 161. Washington, D.C.: GPO.

In addition, the intention of the 1943 act appears to have been to separate telegraph from telephone in the domestic market. The act authorized the merged company "to acquire the domestic telegraph facilities . . . of any carrier which is not primarily a telegraph carrier" (cited in Borchardt 1970, 28). But AT&T was said to have obstructed the sale of its teletypewriter exchange (TWX) service.[28] Although Congress had evidently intended to effect that split, in a tribute to AT&T's growing power, in 1964 the FCC assured the company that such a policy "was not justified and . . . might not serve the public interest" (cited in Borchardt 1970, 28–29). AT&T's 1943 refusal to give up its domestic telex business held long-term implications for FCC regulation of the sector.

The divestment of Western Union's international transmission in the 1943 amendment to the 1934 Communications Act was intended to take place within a "reasonable time" [Sec. 222 (c) (2)], but, due to successive problems, the FCC was obliged to give annual extensions. At one point Cable & Wireless was approached, but the British company was not eager to buy old telegraph cables. Subsequent U.S. proposals, involving all the record companies, then broadened out to an exchange of assets based on British and U.S. spheres of influence. The U.S. companies would have taken British assets in the Caribbean, and there would have been joint ownership in South America. But the British withdrew from negotiations after FCC actions in South America (see below).[29] The sale to Western Union International Inc. (Xerox) was finally concluded on January 15, 1969. It once more emphasized the impact of the 1943 act—that the immediate postwar international telegraph market was based on a state-to-state basis with separation of the domestic and international (except for AT&T).

FCC Unilateral Action: South American Empire

In 1941 the FCC began an examination of telegraph and telephone rates between the United States and South America. Concluding that the earnings of the telegraph companies were too high, in the autumn of 1942 the FCC insisted that RCA and All-America Cables reduce tariffs to South and Central America and the West Indies by 20 percent (FCC 1943, 37 and 40). Fly's aim was to save U.S. customers about $2 million annually. In order to implement those cuts, despite opposition from the companies, the FCC imposed inward and outward telegraph rates in the U.S. license to Columbia.[30]

Then, in 1943, the FCC ruled that the companies should unify certain classes of telegrams. These unified rates were to equal 75 percent of the existing letter rate for commercial telegrams and 50 percent for government traffic. It further ruled that both southbound and northbound rates should be equalized and that settlements should be paid in dollars rather than the gold francs adopted by the ITU. But because they contravened ITU regulations, a number of South and Central American governments refused to accept the FCC's unilateral decisions.

A joint delegation from the FCC and State Department to South American capitals in 1944 failed to persuade some governments to break ITU regulations.[31] In addition, in Brazil, the Western Telegraph Company (owned by Cable & Wireless) refused to implement the unified rate. Cable & Wireless and the British government contended that the FCC had acted "ultra vires in dictating and arbitrarily enforcing the rates to be charged by a foreign company in a foreign country on the ground [sic] that the traffic was ultimately destined to the U.S."[32] With the Brazilian government's agreement, in order to demonstrate that the FCC was interested not in low tariffs but in extending its regulatory power over South America, the company designed new rates. When the FCC rejected these rates, the Brazilian government and the company submitted a further set of rates. These conformed to ITU regulations but were higher than those previously proposed. Again the FCC rejected them.

Meanwhile the reduction in Cable & Wireless's tariffs attracted business from ITT's All-America Cables Inc. In December 1944, on the grounds that it had failed to show the "justness and reasonableness" of the differentials between its northbound and southbound rates, the FCC ordered Western Union to cease handling all except government messages from Brazil to the United States.[33] In reply to Western Union's contention that it was under contract to interconnect with Cable & Wireless, the FCC stated that "the contracts of a carrier cannot constitute a justification for the handling of communications at charges which are unjust or unreasonable, unjustly or unreasonably discriminatory or otherwise unlawful."[34] This edict meant that the Cable & Wireless subsidiary, Western Telegraph, could not transfer traffic to Western Union for delivery within the United States. The position worsened when the FCC blocked the two companies exchanging traffic from Brazil in London or Canada. Cable & Wireless started to lose £120,000 ($480,000) annually.

The British Foreign Office argued that "to admit the right of the FCC so to block traffic would be tantamount to admitting the jurisdiction of the FCC over all foreign as well as domestic operating companies that accept traffic destined for the United States." The office also said, "The action of the FCC in forcing Western Union to break a contract with their partner raises serious issues . . . as in all America . . . no commercial agreement would be secure."[35] It wondered if AT&T's contract for radiotelephone with the British Post Office was safe.

The U.K. government could not protest against the general rate pattern proposed by the FCC. It had not only proposed a similar unification at the 1938 ITU Cairo Conference but had suggested that it would be in order for any two countries or a group of countries to unify their rates reciprocally. In addition, the U.K. government had departed from the International Telegraph Regulations when it introduced the Empire Rate scheme in 1937. Also, since the United States was not a signatory to the International Telegraph Regulations, it was not bound by Article 32, which prescribed the gold franc as the monetary unit for settlements.

Instead, the U.K. Foreign Office took the position that

> the U.S. Government [is] claiming the right to dictate and enforce the rates to be charged by a foreign company in a foreign country for all traffic to destinations in U.S. territory although the Administration of the foreign country concerned has given its approval to the rates of the foreign company and although those rates are in accordance with the international agreements to which both the foreign country and the national Administration of the foreign company are parties.[36]

It objected to the unilateral methods adopted by the FCC, to its dictatorial attitude to the companies, and to its lack of consultation with the British.

A delegation from the FCC, State Department, and industry visited Europe in March 1945 to undertake bilateral consultations on the issues. Discussions between an FCC Commissioner, Ray C. Wakefield (a Republican from California and FCC Commissioner from the end of 1941), and Cable & Wireless produced a compromise on a set of rates from Brazil to New York.[37] A press notice was agreed between the three governments.[38]

The British also thought they had agreed to mutual consultation before any further reductions. Yet, within a month of the U.S. delegation's return, the FCC announced cuts in the rates to and from liberated Italy. The British ambassador in Washington was instructed to inform the State Department that the British government "received the report of United States action in this matter with incredulity."[39]

After a joint meeting with the FCC, U.S. State Department personnel asked the U.K. embassy staff in Washington for help in strengthening its position versus the FCC, with which it had a "strong difference of opinion . . . on the scope of the latter's functions."[40] Then came news that the FCC was attempting to "force" the Brazilian government into repudiation of the rates agreed at the U.K.-U.S. meeting of March 1945. Following representations from the British, the U.S. State Department pointed out to Wakefield that the Commission's denunciation of the London agreement called into question the good faith of the United States. But Brazil had already caved in to FCC demands and there was no going back on the new northbound rates.[41] The whole matter left the State Department highly embarrassed and the British concerned that the State Department was unable to control the FCC. But the FCC had demonstrated its regulatory power over the U.S. companies, had met political New Deal goals by reducing tariffs for U.S. consumers, and had shown that end-to-end international networks could be subject to U.S. domestic regulation. We can infer that for the FCC, domestic political priorities were more important than international opprobrium.

"Freedom of Information" and Liberalization of the British Empire

Toward the end of 1943, James Fly began publicly to link his plan to amalgamate the U.S. international record carriers with access to the British Empire, particularly India and South Africa. Fly teamed up with Kent Cooper (whose book detailing the Associated Press's crusade to break British Reuter's news agency monopoly was published in 1942) and Cranston Williams of the American Newspaper Publishers Association. They demanded that the United States be able to deliver its news and its radio pictures around the world. In 1943, Cooper underlined to Secretary of State Cordell Hull "the imperative necessity to incorporate in the international agreements to be made after the war stipulations regarding freedom of the press throughout the world and equal access by all to the news and the means of transmitting it."[42] Following a grassroots campaign by the American Society of Newspaper Editors, in September 1944 the House and Senate passed a unanimous joint resolution calling for a "world-wide right of exchange of news . . . without discrimination . . . protected by international compact" (Schiller 1979, 348).

In this atmosphere Fly harked back to the post–World War I situation, where the British and French had seized the cut German submarine cables (Hills 2002, 180–82). He argued that the British were still in control of world communications, particularly shortwave radio. Fly cited British refusal of an Australia–United States direct press circuit, of direct communications with South Africa, and of relay stations to pick up and copy U.S. news in India. Perhaps as a bargaining ploy, he suggested that the U.S. military network should be utilized as "an electronic conveyor belt around the equator with transmitting stations to relay messages both North and South to reach all parts of the world." By early 1943 the Army and Air Corps had completed an around-the-world wireless pipeline, extending from Eritrea via New Delhi, Brisbane, and San Francisco to Washington, carrying fifty thousand messages a day. Fly suggested that using the pipeline would save frequencies and avoid atmospheric troubles.[43]

By autumn 1944 Fly's campaign had evolved into five "basic principles of freedom of communication: uniform rates for all messages in all directions; low rates for these messages; instantaneous radio communication to all important areas of the globe; uniform and low press rates throughout the world; freedom of ingress and egress of information in all nations."[44] Based on a postal model, Fly wanted world uniform rates of fifteen cents per word or lower.[45] The U.S. press commented that "beneath the international proposals lies a plan for which Mr. Fly is campaigning to merge all United States companies doing international business."[46]

The British attitude was summarized by an embassy official in Washington: "When he says that he wants 'complete freedom for all peoples of the world to communicate directly with each other' he means, of course, primarily that he wants freedom for American telecommunications agencies to establish direct circuits between the United States of America and any point on the globe."[47] In London, Fly's views were seen as an attempt to advance "American commercial dominance under guise of a democratic philosophy for the free flow of international communications." Reuters argued that the existing two-cent Empire press rate was equally available to British and American press agencies.[48]

By 1944 the U.S. Navy had joined in the clamor for the merger of the international carriers into a company established under congressional charter with the War, Navy, and State Departments represented on the board of directors.[49] ITT was enthusiastic and the other companies generally in favor.[50] But the centralized control envisaged by the Navy brought it into conflict with a State Department that favored competition.[51] AT&T was also hostile to the proposed merger, and Francis de Wolf of the State Department told a British embassy official late in 1944 that "there was no intention of interfering with them [AT&T]."[52] AT&T was too powerful for the FCC.

At this time the British Post Office thought Colonel Sosthenese Behn, head of ITT, was the likely candidate to control the proposed merged company. They thought he wanted markets for his manufacturing interests.[53] The British embassy in Washington explained, "There is an atmosphere of continuous conflict between the FCC and the carriers, the FCC on the whole maintaining a New Deal psychology and adopting an almost crusading policy in favour of lower rates and the interests of consumers generally. Like other New Deal policies, that of the FCC is also strongly tinged with nationalism. The latter brings the FCC into further conflict with the State Department."[54] Meanwhile, the FCC and its chairman were coming under increasing Congressional scrutiny. Fly got into trouble partly because he had refused to accede to the 1940 request from FBI Director J. Edgar Hoover, "to monitor all long-distance calls passing through New York to . . . territories occupied by German and Italian forces" (Edwardson 1999). The dispute brought Fly into decades of conflict with Hoover.

In January 1943 a House resolution authorized an investigation into the FCC. Fuelled by a dispute over a radio station license, Rep. Eugene G. Cox, Democrat from Georgia, attacked both the FCC and Fly as Communists and "the nastiest nest of rats to be found in the country."[55] In the House investigation, a secret 1942 memorandum written by Adm. Stanford C. Hooper became public. He accused Fly of preventing wire-tapping legislation that would have averted Pearl Harbor (Edwardson 1999). Fly then accused Hooper, now retired, of seeking to replace him as chairman of the FCC and denied that he had been instrumental in having Hooper retired.[56] Nevertheless, Fly resigned later in 1944, to be replaced by Paul Porter. From 1946, Fly continued his antiwiretapping campaign as director of the American Civil Liberties Union. His "freedom of information" campaign took on a life of its own and, decoupled from the telecommunications issue, bore fruit in Article 19 of the U.N. Universal Declaration of Human Rights of 1948.

Franklin D. Roosevelt died April 12, 1945, and was succeeded by President Harry Truman. In 1946 the Republicans won House and Senate majorities, and the merger of international carriers met with increasing congressional approval.[57] But the FCC's international cartel and wartime traffic had ensured that all the international record companies were profitable. Only ITT wanted a merger. In February 1947, the Board of War Communications was abolished and the FCC lost its wartime powers.[58]

Bilateral—Bermuda Agreements 1945

The British delayed a bilateral meeting requested by Fly's FCC until after Cable & Wireless's future had been decided. Eventually formal intergovernmental discussions between the United Kingdom and Dominion governments and the United States took place in Bermuda in late November 1945. The U.S. delegation included Paul A. Porter, Fly's successor. Seventeen representatives of private U.S. companies, together with six from commonwealth companies, also attended.[59] The U.K. delegation hoped to benefit from the known dissension between the U.S. participants, but Chairman Porter dominated the proceedings. According to the British delegation, he began with a series of peremptory demands that were similar to those of Chairman Fly, but, when these were rejected, the meeting became friendly.

In the short term the U.K. government was primarily concerned with preventing the postwar retention of all the direct U.S.-commonwealth circuits and their use for transit traffic. It had devised a traffic formula to determine whether a circuit should be retained, but it compromised on nine circuits—some, such as the U.S.-India route, where traffic was low. The U.S. delegation wanted unrestricted transit traffic on the circuits but acceded to British opposition.[60] And although the FCC preferred a twenty-cents-per-word tariff to Europe, the meeting agreed that a tariff ceiling of thirty cents (1s/6d) would apply both between the United States and the British Commonwealth and between the United States and countries outside the commonwealth. The new ceiling of thirty cents involved U.S. companies in reducing rates, often by more than 50 percent. The agreement also provided for mutual consultation prior to future changes in international tariffs.[61]

Back home the FCC then took the opportunity to expand the concept of the "international formula." In bidding for the remaining radiotelegraph circuits to the Commonwealth, each of the companies was allowed "to obtain as much [Commonwealth] traffic proportionately as they enjoy on a worldwide basis."[62] RCA was allocated 63 percent and Mackay 37 percent (Oslund 1977, 189n43). Mackay, the loser in the share-out, was to challenge the "single-circuit policy" for the next ten years, eventually overturning it in 1956.[63]

The Bermuda Agreement soon ran into trouble. Under the Defense Board edict of 1943, the FCC had to issue a license before a company could negotiate with a foreign government. In December 1945, on entering into negotiations with the Netherlands for a reduction in tariffs, RCA found that the commission had already written a twenty-cent collection charge into its landing license. Despite company protests, FCC Chairman Porter instructed the company to proceed.[64] Then, without notifying the British of the changes, in May 1946 the FCC extended the twenty-cent rate unilaterally so that the affected routes had differential customer tariffs at either end.

The British government made strong representations "against this unilateral procedure, stating that . . . the Bermuda Agreement . . . had specifically stated that the thirty-cent collection charge should be offered to other countries."[65] In subsequent informal discussions in Washington, the British commented that "the State Department representatives present seem extremely afraid of the FCC whose nationalism they obviously deplore but they do not seem in a position to do anything about it."[66] Less than one month later the British complained that the FCC had reduced rates on other routes without consultation and was delaying consideration of their applications for tariff alterations. In general the Post Office felt that it was losing from the Bermuda Agreement.[67]

The Death of the Bermuda Agreement

Because of deteriorating economic conditions in Europe and Asia, four of the five major U.S. carriers expected substantial deficits under the thirty-cent rate.[68] Similarly, Cable & Wireless estimated a loss of £1 million due to the Bermuda Agreement.[69] Then, hit by increased wage rates, Western Union filed for an increase in domestic tariffs to make good a deficit of $19,251,000 in its 1946 operations.[70]

During this period Western Union and Commercial Cable were still running their end-to-end telegraph services into Britain and operating all but two of the transatlantic cables.[71] In the long term the British wanted to get rid of them. Behind this desire lay a resolution of the Australian 1942 Commonwealth Conference (passed with U.S. end-to-end radio networks in mind) that all terminals in territories of partner countries should be owned, operated, and controlled by British interests. To its embarrassment the U.K. government had had to enter a reservation to the commonwealth resolution.

Western Union's financial predicament helped in preventing a rates war. In 1949 a revised Bermuda Agreement raised rates and for the first time acknowledged that radio should not compete to the detriment of cables. A further revision in October 1952 increased the maximum for collection charges once again and improved the British position by a reduction in the dollar/sterling exchange rate on which they were based.[72] Once more the U.S. delegation proposed that the U.S. radio circuits take transit traffic, and once more the British deferred for commonwealth agreement. The British also began freezing out the U.S. telegraph companies from Britain by refusing to raise the loss-making Westbound transatlantic rate.[73] The British cabinet relented on tariff increases only after TAT-1 (see p. 57) had come into operation and the U.S. end-to-end networks became redundant.

Bringing the ITU under U.S. Control

In October 1942 Germany brought into existence a European Telecommunications Union consisting of Germany, Italy, and eleven other satellite or occupied countries of central Europe. The Union's territory was to be treated as one state with a uniform tariff and no accounting charges between countries (Mance 1943, 7 and 14). Possibly in response, in the summer of 1943 President Roosevelt showed Prime Minister Churchill a secret letter from Cordell Hull enclosing a report from a Special Interdepartmental Committee on Communications setting out the terms of peace to be imposed on enemy nations. The plans were unlike those of the 1919 Treaty of Versailles that had enforced adherence to the then ITU Convention and Regulations (Mance 1943, 71).

The Committee envisaged that an international telecommunications organization would be "centralized in the world political organization" (what we now call the United Nations) in which the Allies with the largest telecommunications interests would have veto power. The enemy nations would have to agree to abide by its "regulations, directions and orders." Further, the Axis states would have to guarantee foreign companies "the unrestricted right to send and receive international correspondence . . . under the same service and tariff regulation as apply to that enemy government and its nationals respectively"—a "national treatment" provision that could ensure end-to-end working for U.S. companies. The Axis also had to guarantee "the secrecy of all international correspondence" and guarantee their nationals freedom to listen to radio broadcasts and receive communications from foreigners—freedom of information provisions that were said to emanate from Chairman Fly.[74]

The plan included elements—such as the Allies' veto power—that copied similar proposals in 1920 that foundered on the opposition of the U.S. companies (Hills 2002, 184). The British were unenthusiastic, and at some point the idea of this new postwar institution was replaced by that of revamping the existing ITU within the United Nations structure. Perhaps the change was due to a 1943 report by Sir Oswald Mance, later appointed Chair of the Temporary Transport and Communications Commission of the United Nations Economic and Social Council. He suggested a future structure for the ITU and by 1945 the U.S. government had put forward very similar proposals. The Berne Bureau was to function as a secretariat to the ITU; there would also be a permanent administrative council and a Central Frequency Registration Board to improve registration of radio frequencies. Also, a Rates and Traffic Board (a U.S. innovation) would provide a "forum for the solution of rates problems and settlement of international disputes."[75] The United States also proposed modification of the scale of contributions and that English should join French as an official language.

In general, the British agreed with U.S. proposals but opposed all those that might place the ITU above sovereign states. Also, they could see no purpose in a Rates and Traffic Board staffed by permanent officials.[76] Opposition from the U.S. international record companies ensured the Rates Board plan was dropped, but there is no clear indication of the motivation behind the proposal.[77] It may have been designed as a means of bypassing U.S. companies' opposition to those ITU regulations that specified tariff structure. It may have been designed to promote AT&T's 1944 World Plan.

AT&T's plan "applied specified distance-based accounting rates ranging from $US 1.50 (4.5 gold francs) for direct links of less than 500 miles to $US 4.0 (12 gold francs) per minute for intercontinental telephone links over 3000 miles." The plan was based on AT&T's bilateral arrangements for telephone links with other countries in which the rates were the same in both directions and the charges divided equally between the parties. It formed the basis for postwar telephone "accounting rates" (Ergas, Paterson, and Geissler 1989, 5).

Throughout this period the British were concerned at the growing isolationism of the FCC. A senior official wrote that the FCC

> [has] recently become more and more dominant as the American governmental authority in this sphere. The State Department gives the impression of doing a rather ineffective best to mediate. (State Department is not on good terms with FCC, nor are the U.S. private companies). I have little doubt that were it not for the desire of the FCC to make home political capital out of telecommunications issues we should by now have made considerable progress towards fuller American participation [in the ITU]. Mr. Fly (ex-Chairman of FCC) made public anti-British pronouncements; his successor, Mr. Paul Porter, refrained from this but his activities were anything but collaborative.[78]

British concerns were split. They wanted to prevent the United States from moving the ITU from Berne to New York, where U.S. radio manufacturers would gain increased influence. They also aimed to prevent the United Nations (then dominated by the United States) from gaining regulatory power over the ITU.[79] But the stable postwar system they so desired was only possible if the United States became a full member of the ITU. And, although they were aware that European governments resented the U.S. attitude toward the ITU, they knew that Europe had "nothing to offer the Americans in return for what would be for them a difficult concession to make."[80]

It is not clear why the Truman administration decided in favor of joining the ITU. There is some evidence that it saw the ITU as a means of strengthening the United Nations and perhaps feared that without internal pressure from the United States, the Europeans might not have linked the two. In 1946 the Temporary Transport and Communications Commission of the

United Nations Economic and Social Council reported on the future of the ITU. Its report reflected the views of its British chair, Oswald Mance, rather than those of George Baker of the State Department.[81] It recommended, among other things, that the ITU charter allow it complete autonomy from the United Nations (Mance 1946, appendix IV chart).

Reflecting President Roosevelt's courtship of Stalin and the generally poor relations between the United Kingdom and the United States, the British Post Office attempted to bring the postwar ITU conferences to Europe. But the U.S. State Department preempted these British maneuvers through the U.N. Communications Commission.[82] The debate in that commission seemed to confirm Post Office fears. It made it "quite plain that the object of the United States in forcing the conference to be held there is, as Mr. Gross has been saying in Europe for some months past, to remould the Union in a new cast, and that in future it will be more amenable to political pressure from the United States designed to satisfy their domestic pressure groups interested in telecommunications (mainly radio)."[83] The British perceived the Soviet Union's backing of the United States as an attempt to gain "individual representation with voting power of as many of the Soviet republics as possible in the various organizations associated with the United Nations."[84]

The postwar ITU conferences were to begin in May 1947 in Atlantic City, New Jersey, with a preliminary conference held in Moscow in 1946 to set an agenda. The subsequent control exercised by the U.S. government mirrored their control over the 1927 Washington Radiotelegraph Conference that followed World War I (Hills 2002, 202–4). A joint U.S. industry-government group began meeting in December 1946 and twenty-four U.S. operating companies or groups sent 105 representatives to Atlantic City. Playing to British fears, the U.S. delegation was headed not by the State Department but by the new chair of the FCC, Charles Denny Jr.

Atlantic City Conferences

The Atlantic City meeting originally comprised three conferences in succession: the International Radio Conference called to revise the 1938 Cairo Radio Regulations; the plenipotentiary of the International Telecommunication Union to revise the 1932 Madrid Convention; and the International Conference on High Frequency Broadcasting that was subsequently postponed.

In general, the International Radio Conference accepted U.S. proposals, such as the establishment of an International Frequency Registration Board (IFRB) and (over British opposition) the reworking of the whole frequency allocation.[85] The United States was also successful in replacing the loosely structured prewar organization with a stronger organization independent of the Swiss government and linked with the United Nations in accordance with Article 57 of the U.N. Charter. The ITU gained an administrative council, a secretary general, and two assistants. The technical committees—on telegraph (CCIT), telephony (CCIF) and radio (CCIR)—were brought into the organization, each with its own director.

The most contentious issue surrounded voting rights. At Stalin's request the United States successfully proposed the exclusion of Franco's Spain from the conference. Anxious to prevent the Soviet Union's satellites from gaining votes, the British successfully proposed limiting voting rights to sovereign states and associate membership without voting rights for nonsovereign territories (colonies). But substantial exceptions had to be made to satisfy the European colonial powers and it was not until 1973 that the practice of "colonial" voting (colonizers voting for their colonies) within the ITU ended.

Interestingly, in a move to be repeated in the 1990s, the U.S. government proposed alterations to the ITU convention that limited the autonomy of companies. Whereas from the beginnings of the ITU private companies had been able to have direct relations with Berne and since 1875 had had a right to attend ITU conferences, the U.S. proposal subordinated them to governments. Under the 1947 convention private operating agencies could gain entry to plenipotentiary conferences and administrative conferences only as part of their member country's delegation (Feldman 1975, 92). There could not be a rerun of the 1932 Madrid and the 1938 Cairo conferences when Western Union's foreign-born employees were refused entry to the U.S. delegation but took their place with the British.[86]

For the first time, in recognition of the U.S. system, the convention and regulations of 1947 mentioned "recognized private operating agencies" (RPOAs). The Atlantic City convention article on "special arrangements" stated:

> Members and Associate Members reserve for themselves, for the private operating agencies recognized by them and for other agencies duly authorized to do so, the right to make special arrangements on telecommunication matters which do not concern Members and Associate members in general. Such arrangements, however, shall not be in conflict with the terms of this Convention or of the Regulations annexed thereto, so far as concerns the harmful interference which their operation might be likely to cause to the radio services of other countries (ITU 1947, Chapter III, Article 40).

From the U.S. perspective, the inclusion of RPOAs under these articles meant that where its companies operated end-to-end systems, they were not subject to ITU regulations.

In general the British considered the plenipotentiary to be successful and it signed the agreement on behalf of itself and what became known as the "Colonial Ensemble." They had defeated the U.S. proposal that the ITU should be sited with the U.N. in New York: it would now be located in Geneva. The U.S. manufacturing companies had not gained overt influence. The British had also defeated proposals designed to forward the political objectives of the Soviet Union. English was now an official language (although at the expense of agreeing to Spanish) and overall the British hoped for "a greater and more balanced participation of the U.S. in activities of the Union . . . in the future."[87]

Revising the ITU's Telegraph and Telephone Regulations

The fourth ITU conference—the Telephone and Telegraph Administrative Conference in Paris in 1949—presented the U.S. government with a problem. It had never been a party to the International Telegraph Regulations, but the 1947 Atlantic City convention made the four sets of regulations (telegraph, telephone, general radio, and additional radio) binding on all members of the union. The Paris conference was expected to agree to changes that would make the regulations acceptable to all.

In general the U.S. companies were opposed to their government's signing the regulations. Prior to 1934 the United States had sided with the companies and argued that it did not have the power to transpose the international regulations into domestic regulation. However, after the creation of the FCC, it was not only difficult to argue this case, but the agency had an interest in expanding its regulatory power. At the Atlantic City plenipotentiary, eight countries—including the United States—signed a protocol agreeing to bind themselves to the Telegraph and Telephone Regulations after the Paris revisions.[88]

For the United States to sign them, the ITU European Telephone Regulations needed their geographical remit extended. Although AT&T objected to any international regulation, it allowed that "should it become advisable for this Government to adhere to some form of international telephone regulations, the AT&T would not object to adherence by the United States to such regulations, provided they were of such a general nature as to leave American industry free to carry on their operations under their own standards." The arrogance of the company's words is breathtaking, even fifty years on, but they indicate AT&T's domestic and international power. The company provided the only U.S. representation at the ITU International Telephony Consultative Committee (CCIF) in June 1948.[89]

The ITU Telegraph Regulations were in need of revision. By 1947 fourteen members had failed to approve the 1938 Cairo Revisions of the 1932 Madrid Telegraph Regulations. In addition, fifty-eight countries, including the United Kingdom, had entered reservations on the rule setting local currency rates—a rule that meant when a country devalued its currency, it had to increase domestic charges.[90]

One problem for the United States was the method of calculating international tariffs. The U.S. record companies objected to the ITU method of "terminal charge" and "transit charge" that were roughly based on "costs." The delegation argued that "it cannot agree to provisions . . . establishing terminal and transit rates for the United States."[91] Tariff setting in the United States depended on what the market would bear, rather than on the "costs" of operators. The U.S. system was designed to block competitors and allow profit-making over cartelized routes, whereas the ITU method was designed for interconnectivity. The U.S. official objections were mainly that the regulations breached company autonomy.

For the U.S. companies there was also a problem that the regulations demanded that charges be equalized in both directions.[92] Despite FCC public statements on equalization of tariffs, in practice the record companies made up their losses on traffic leaving the United States with profits in the other direction. For instance, on the U.S.-France route, the U.S. companies (with their own offices in France) collected twenty gold centimes more in France than they collected in the United States. France was losing exchange to the United States and effectively subsidizing U.S. domestic users.[93]

Most U.S. attention at the ITU Regulations Revision Committee in January 1949 went to Article 40 of the convention, which allowed "special arrangements." The question was how these "special arrangements" would be interpreted in the regulations. First, the U.S. delegation attempted to exempt the U.S. telegraph companies from all ITU regulation by dint of rewriting article 1 that defined the scope of the regulations. The United States proposed that the regulations should not apply to private or regional agreements or to services run by one private operating agency. The intention of the amendments seems to have been to exclude pan-American networks, such as Western Union's expansion into Canada and Mexico, and second (since Western Union and Commercial Cable already worked to ITU regulations in Europe) those of ITT's end-to-end services in South America. As the French pointed out, the United States was attempting to gain a definition of the exact meaning of the phrase "special arrangements."

Although at first demanding that there should be no regulation of a company's end-to-end services, the U.S. delegate eventually accepted that there could be foreign regulation but demanded that each private operating agency "must be free to operate its daily routine as it saw fit, meanwhile being subject to regulations of the Administrations in the countries in which they [sic] operated regarding rates and certain other matters."[94] In the face of reassurances that private companies were not subject to political interference, having successfully expanded the article to include regional agreements, the United States withdrew the amendment designed to exempt individual companies from ITU regulations. However, the U.S. delegation repeated the same arguments at the full plenipotentiary.[95] Yet ironically, at this time the FCC refused to allow the French Cable Company to become a party to the "international formula" or expand its operations in the United States until the French government severed its ownership of the company. It allowed that expansion only in 1979.[96]

Despite heated objections from U.S. newspapers prior to the conference, at the Paris 1949 Conference the U.S. delegation pressed to unify tariffs and abolish deferred telegrams—in effect legitimizing its unilateral action in South America and ensuring the compliance of Cable & Wireless. Within the United States the very act of negotiating international regulations increased the power of the FCC. It demanded detailed operational data from international record companies in order to support its negotiating position.[97] In addition to those made in Paris, further U.S. reservations to the Telegraph Regulations came out of consultations with the record companies and large users.[98] In content there was not much left to apply—in 1949 U.S. reservations to the Telegraph Regulations covered 34 articles—but the FCC had institutionalized its position and gained domestic power (U.S. Congress, Senate 1950).

In terms of the telephone regulations, at the Paris meeting the U.S. delegation followed the views of AT&T in arguing that the European Telephone Regulations should not be extended to the rest of the world. The FCC delegate insisted that the broad features of the regulations were already covered in the 1947 ITU Convention and that the United States had belonged since 1925 to the CCIF. Then, when the Paris committee wanted to refer the matter to the CCIF for a report to the 1952 Buenos Aires plenipotentiary, the United States refused to cooperate, insisting that the principle of extension should be discussed first.[99] The 1949 regulations continued to apply only to the European system (ITU 1949, Article 1(1)).

Subsequently, the 1952 plenipotentiary revisited the 1947 convention and decided that rather than binding all members to the regulations, the ITU should adopt a "more flexible approach" (Codding and Rutkowski 1982, 214). In signing the 1952 convention the United States refused any obligation with respect to the Telephone Regulations or the Additional Radio Regulations (ITU 1953, 112). But automatic technology was increasing the need for world-wide standardization of operation. Countries in the Far East, Middle East, and South America asked the CCITT (amalgamated from the CCIT and CCIF in 1956) to create universal telephone regulations. In 1958 the administrative conference then simplified the telephone regulations and extended them worldwide on a permissive basis. But the United States refused to sign them on the grounds that they were incompatible with the U.S. system and would freeze the technology (ITU 1959, 57–59). It was the only ITU member not to sign.

Although the 1965 plenipotentiary reversed the 1952 decision and made signature of the regulations compulsory, it took until 1973 before the United States signed. Even then it exempted both telegraph and telephone service with Canada and Mexico and (unnecessarily) the application of the regulations to private lines (ITU 1973, 78). The formal explanation for the U.S. signature is that the 1973 regulations were further simplified (Codding and Rutkowski 1982, 37 and 53). However, a cynic might suggest that in the 1970s AT&T's national monopoly was under attack, and ITU international regulations provided AT&T with a legal mechanism against IBM's global pretensions.

The Military-Industrial Linkage: ITT

In the 1950s, with the cold war gaining momentum and the Korean War pitting the United States against China, the military gained ascendancy within the U.S. economy. In his farewell address in 1961 President Eisenhower warned against the "military-industrial complex," of which the communications industries were part.[100] Nowhere was this relationship better illustrated than in the case of ITT.

For a company with factories spread throughout Europe, the war could have been disastrous. But ITT's European factories flourished, supplying both Axis and Allies (Sobel 1982, 110). Its Argentinean telephone company—the United River Plate Company—also extended its pro-Axis influence throughout Latin America. However, by 1943, after an FCC report, U.S. agents were placed in ITT offices in Bolivia, Paraguay, and Spain (Sampson 1973, 41). Also, congressional suspicions prevented ITT from buying Western Union's cables.

Nevertheless, at the end of the war Colonel Behn once more forged close links with the Pentagon. He accompanied U.S. troops on their entry into Paris and, supported by Jame Forrestal, then Secretary of War, unsuccessfully petitioned the Truman administration for financial help to buy ITT's rival in Mexico owned by Swedish L. M. Ericsson. Behn argued that Ericsson was from a neutral country and therefore unreliable (Sobel 1982, 191; Sampson 1973, 44n).

Meanwhile the FCC's "international formula" effectively prevented the international telegraph companies from expanding market share unless they expanded their named company traffic. One potential way of doing so was to provide quicker transmission over new facilities. But by 1952 the technology of underwater repeaters developed by AT&T and the British Post Office had progressed sufficiently for a transatlantic coaxial telephone cable, TAT-1, to be negotiated between Britain, Canada, and AT&T. Coaxial cable technology allowed transmission of both telegraph and telephone signals. The FCC allowed a 1950 coaxial cable to Havana shared between AT&T and ITT offering both services, thereby threatening AT&T's entry into the international telegraph market.

ITT responded to this threat by forging an alliance with the Pentagon to propose a new coaxial cable running from Massachusetts to Scotland via Newfoundland, Greenland, and Iceland.[101] Called "Project Deep Freeze," the cable would provide 120 leased-line telegraph channels. The company and the U.S. State Department claimed that the cable would provide facilities for the North Atlantic Treaty Organization (NATO) and "an urgent requirement of the [U.S. Air Force] for interruption-free communications service" between countries whose proximity to the Soviet Union had increased their security importance.[102] But Pentagon take-up on the cable was minor.

Not only did the Post Office not believe ITT's proposed use of the cable, but the prospect of a new transatlantic telegraph cable with a capacity greater than all that was then currently available conflicted with its policy to get rid of the U.S. record companies from the United Kingdom. So enraged was the Post Office at the Pentagon's subsidy to the company that the British government sent a diplomatic note demanding both reciprocity of entry to the United States and permission to run telegraph circuits over TAT-1. The answer was negative: the U.S. government supported the "Western Union" model and separate international telegraph and telephone facilities.

Finally, after increased pressure from the Eisenhower government, in February 1955 the British agreed to a landing license for defense purposes only. Dissatisfied with this compromise, the State Department orally suggested a traffic cartel—a revamping of the "international formula" to include the British Post Office—or joint ownership of the cable. But to the British the implication of a license for ITT was that all the foreign record companies in the U.K. could demand new licenses.

Nor did talks in Washington help. The record carriers had already filed objections to AT&T offering telegraph facilities over its radio link to the Canadian border for the new transatlantic telephone and over its projected coaxial cable to Hawaii. AT&T refused to support British demands that TAT-1 carry both voice and record traffic. The British delegation concluded that "the communication operating companies are extremely powerful and are strongly against any form of Government control. This . . . is taken more strongly by the American Telephone and Telegraph Company than by any other company. It is illustrated by the fact that in our talks the United States Government agencies were not allowed to state their views until prepared statements of these views had been vetted by the companies (we believe word by word)."[103]

However, in private conversation, Commissioner Webster explained that the FCC was embarrassed by U.S. government support of Project Deep Freeze. He disclosed that a recent defense review had altered its relevance so that "the importance of the project from the defence point of view was now probably insufficient to outweigh its commercial disadvantages in the sense that it would disrupt the competitive balance among the overseas tele-graph carriers."[104] Finally, the U.S. side agreed that it would look at whether the companies could operate competitively at the same time as jointly own-ing circuits in a coaxial cable. For its part, the British government agreed to reconsider the landing licenses and to reconsider Project Deep Freeze.[105]

It rejected Project Deep Freeze but granted twenty-five-year licenses to the U.S. record companies, limiting them to existing services and technology. The companies were not allowed to provide telex or to replace their exist-ing cables.[106] After ITT's further representations to the State Department failed, the British formally confirmed their decision on Project Deep Freeze in September 1956. Although in 1957 ITT attempted to revive it as a purely defense project, and the company's lobbying went to the highest levels, the success of TAT-1 spelled the end for separate telegraph cables (Sobel 1982, 63). In 1959, after Harold Geneen became the company's CEO, ITT launched a lawsuit against the U.S. government for lack of support in its negotiations with the U.K. government.

The Impact of TAT-1

The first transatlantic coaxial cable, TAT-1, provided thirty U.K.-U.S. telephone circuits together with six telephone and eleven telegraph circuits to Canada. Costing £120 million ($334 million), it came into operation on September 25, 1956. During its first year of service it carried 220,000 calls between Britain and the United States and 75,000 between Britain and Canada, bringing in £2 million ($5.6 million) in revenue for the three operators.[107] So successful was it that AT&T brought forward to 1959 another cable, TAT-2, with forty-eight voice circuits from the United States to France.

The success of TAT-1 turned British thoughts to a direct U.K.-Canada cable. Because the FCC had refused to allow TAT-1 to be used for telegraph to the United States, the eleven telegraph channels on TAT-1 to Canada were almost fully loaded. Post Office telegraph traffic to the United States went over these, then via Western Union's landline. When AT&T decided to bring forward the TAT-2 cable, the company suggested that it could incorporate spurs to the United Kingdom and Canada.[108] It invited the British to take shares in the new cable. But, for the Canadians, the new cable would have meant increasing reliance on the United States for its communications; for the British, a TAT-2 spur would have postponed the integration of transatlantic telegraph and telephone and increased the importance of the U.S. international record carriers.[109]

Instead, despite concern about the financial viability of the project, the 1958 Commonwealth Telecommunications Conference approved a Commonwealth Round the World cable. The first stage, CANTAT-1, between the United Kingdom and Canada came into operation in 1961. The cable, costing $8.5 million, had 70 percent greater capacity than TAT-1 (Brockbank 1958). In the meantime, when TAT-2 was delayed and AT&T needed to lay the cable over the shorter transatlantic route from Canada, its Canadian landing license forbade it from taking direct European-Canadian traffic.[110] In effect, CANTAT-1 bypassed AT&T, the international record carriers' "Western Union" system, and FCC regulation. Its creation brought forward the metamorphosis of transatlantic telegraph communications to a state-to-state basis.

Because TAT-1's technology provided a much faster service than the old telegraph cables, the Pentagon immediately wanted access to both voice and data services. Using the increased capacity of TAT-2 as its justification, the FCC allowed AT&T to breach the divide between domestic and international telegraph. It also authorized AT&T to lease facilities to the telegraph carriers to enable them to provide similar services to the defense agencies (FCC

1959, 119). This was the first instance of international leased lines for data transmission over telephone cables and use of the term "carrier's carrier" to denote the new relationship between AT&T and the record carriers. The new FCC policy also pushed the availability of international and domestic leased lines in Europe up the agenda.

At this time the first cracks in AT&T's domestic monopoly were appearing. A 1949 antitrust suit against AT&T ended in 1956, preventing it from entering IBM's computer equipment and data transmission territory (Olufs 1999, 43). Also, a new technology—microwave transmission—developed in wartime by the Army, offered the potential for fast, cheap construction of transmission paths. In its *Above 890* decision of 1959, the FCC bowed to the lobby of large business users led by Motorola and allowed large companies to construct private lines across long distances.[111] By the end of the 1970s there were more than one thousand private microwave networks over 266,000 miles (Schiller 1982). In turn, AT&T's competitive response began a new phase in domestic regulation. Although the FCC remained accused of "negotiating" rather than regulating AT&T's tariffs, 1959 was the first year when "costs" and formal hearings became part of FCC regulation.[112]

Just as in the 1850s, when Cyrus Field had looked to the international market to give him competitive advantage at home, so now did AT&T. It looked to expand its domestic telex service for business users to the international arena. In 1960 the FCC responded by allowing AT&T to provide nonvoice service to customers other than the defense agencies. Despite Mackay's victory in 1956 over the single-circuit radio policy, lack of shortwave frequencies meant that the record carriers were unable to compete. Only at this point, when CANTAT-1 threatened to bypass its regulatory divisions, did the FCC allow the four record carriers to lease channels on both the CANTAT and TAT cables (FCC 1960, 105). The British had won. Transatlantic telegraph carriage became a state-to-state service. Commercial Cable closed down its end-to-end service into the United Kingdom in 1962 and, having bought out Western Union's international business, Western Union Inc. (Xerox) leased circuits in coaxial telephone cables from 1966.[113]

But the change in FCC policy meant that AT&T was now not only the record carriers' wholesaler but also their retail competitor. AT&T opened TAT-3 in 1963, carrying 138 voice circuits. The following year, ITT's lobbying of Congress led to accusations that AT&T (also then developing satellites) was attempting to create a monopoly in international communications. ITT proposed itself to build the fourth transatlantic cable. This time the FCC responded by limiting AT&T to international voice service, other than for existing customers, and allowed the record carriers to own shares in new cables.[114] International (but not domestic) voice and record carriage were once more separated and the international record companies became state-to-state operators.

Shortly after the decision to abandon Project Deep Freeze, ITT had begun an unsuccessful lobbying campaign for a change in the "international formula" (FCC 1959, 123). In 1964 it filed formally for the formula's revision on the grounds that "the harder a given international carrier works to have messages routed over its facilities by sender's 'Via', the less likely it is to receive a share of the unrouted traffic."[115] Both RCA and ITT accused each other of using loopholes to gain outbound traffic. They demanded direct access to customers—a reversion back to the prewar system of multiple international gateways. But it took until 1980—sixteen years and three chairmen later—for the FCC to reach a decision.[116] In the meantime, the FCC supported AT&T and the state-to-state model of the international network.

Foreign Investment and Dollar Diplomacy

End-to-end networks demanded operators be able to invest overseas. During the interwar period, under Sosthenes Behn, ITT built up a portfolio of national operating companies vertically integrated with its equipment manufacturers and linked together end-to-end with its international lines. Just prior to Pearl Harbor the U.S. State Department helped Behn sell its operator to the Romanian government (Sobel 1982, 100). Then, when postwar nationalism threatened ITT's holdings, the Roosevelt administration helped the sale of its Spanish and Argentinean operators.[117] Behn gradually stripped the assets of the rest and when, in 1959, Fidel Castro responded with nationalization of the Cuban operator, President Eisenhower refused to support the company (Sobel 1982, 304). However, Congress later produced the 1962 Hickenlooper Amendment enabling the U.S. government to cut off aid to any government expropriating without compensation.

On the basis that it was deficit ridden and poorly managed, in 1960 the Brazilian state of Rio Grande do Sul attempted to nationalize an ITT operator for the token sum of $400,000. Harold Geneen, Behn's successor, with State Department support, eventually extracted $7.3 million, half remitted to New York and half paid to ITT's Brazilian manufacturing subsidiary, which also received major equipment contracts (Sobel 1982, 192). In 1967 Brazil paid ITT a further $12.2 million.

ITT had similar success in 1968 when the Peruvian revolutionary military government acquired ITT's 60 percent holding in Compañia Peruana de Teléfonos (CPT) for $17.9 million. ITT was required to reinvest $8 million in Peru. Part of the payment comprised prime real estate and led to a Sheraton hotel in Lima. After the Peruvian company's nationalization, ITT operated only in Puerto Rico, the Virgin Islands, and Chile. Of these, Chile was the largest operation, employing six thousand nationals (Sampson 1973, 257).

Recently declassified documents show that the CIA bankrolled Eduardo Frei, the 1962 Christian Democrat winner of the 1964 Chilean presidential election against Salvador Allende.[118] The CIA director at the time was John McCone. According to the U.S. Ambassador to Chile at the time, Edward Korry, President Kennedy, and Attorney General Robert Kennedy "cajoled U.S. multinationals into pouring $2 billion into Chile . . . for which the U.S. government would arrange guarantees and insurance." They bought up "more than 85 percent of Chile's hard currency earning industries" (cited in Palast 1998). Frei's government awarded ITT a $186 million contract to extend the telephone system, despite a lower bid from Swedish Ericsson. Frei also agreed with ITT on a gradual nationalization of Compania de Telefonos de Chile (Chitelco). By 1970 ITT's interests in the company had been reduced to 70 percent, worth $153 million (Sobel 1982, 302–4).

In 1970, when Salvador Allende narrowly won the Chilean presidency, because ITT was illegally channeling funds into Republican Party coffers, President Nixon could not ignore the company. McCone, by then an ITT board member, is said to have pledged Henry Kissinger $1 million in support of CIA action to prevent Allende from taking office. But, when the CIA murder of prodemocracy Army Commander in Chief, General René Schneider, failed to prevent Allende's confirmation, negotiations on nationalization of ITT's assets proceeded. However, according to Korry, "They [U.S. corporations] pushed the White House to impose a clandestine embargo on Chile's economy . . . ITT paid $500,000 to . . . Jacobo Schaulson, Allende's ally" on a committee set up to compensate firms whose property had been expropriated. Learning of this corruption, Allende refused the compensation and in September 1971 expropriated ITT's properties. Korry alleged, "It was this—the Chilean leader's failure to pay, not his perceived allegiance to the hammer and sickle [Communism]—that sealed his fate" (cited in Palast 1998).

On Korry's return to the United States in October 1971, he advised the Overseas Private Investment Corporation to deny ITT compensation for its seized property and recommended criminal charges against Harold Geneen for falsifying insurance claims and lying to Congress. But although ITT was at first refused compensation and two company executives arrested, in 1974 the Pinochet junta agreed to pay ITT $125 million, and, because they had worked in cooperation with the CIA, the executives went free.

Rumors of the involvement of U.S. corporations in the Chilean coup helped fuel suspicion of foreign investment among Latin American countries and gave rise to the "dependency" ideology of the 1970s, with its emphasis on "cultural imperialism" and autarky (see Bodenheimer 1971). ITT's experience with expropriation and its demands for U.S. government guarantees for private investments in South America also entered the annals of U.S. business history. When, in the 1980s, U.S. corporations began to look to investment in developing countries, they demanded international rules giving them rights that led to the GATT negotiations discussed in chapter 6. Subsequently, in the 1980s ITT sold its remaining European telecommunications manufacturing factories to Alcatel of France and became a finance and hotel business.

Conclusion

This chapter demonstrates how U.S. relational and structural power in the immediate postwar period was limited by the countervailing British Commonwealth bloc of the British and European members of the ITU. Nevertheless, the FCC was able to prevent the British and French from supplying direct telegraph services to the United States and operating within it, at the same time ensuring that U.S. companies could operate end-to-end to Europe. When the FCC's direct power over Cable & Wireless failed, it was able indirectly to exert power by unilaterally revoking Western Union's interconnection contract. It was also able to persuade the ITU that end-to-end working of U.S. companies should not be subject to international regulation and to legitimize its previous unilateral actions on tariffs. Also, Chairman Fly's use of the concept of "freedom of information" to hide the commercial ends of liberalization of the British Empire subsequently became a populist mantra to prevent foreign domestic regulation against U.S. interests. Throughout the period it was domestic interests that fuelled FCC incursions into others' sovereignty.

The separation of the prewar U.S. system from that of the ITU reflected the power of the U.S. companies and their antipathy to the domestic implementation of international regulation. Wartime powers and James Fly together saw the FCC flexing its muscles over the record companies. The "international formula" that froze the U.S. international market in its 1942 structure and the desire of the wireless companies to break into the British Empire increased FCC domestic power. There is no doubt that the FCC could have ignored its bilateral agreement with the British and opened U.S. radio circuits to transit traffic. Perhaps conflicts with the State Department restrained it. And technology may have come to the aid of the British in terms of static interference in radio transmissions from the United States.

The FCC considered it had the right to regulate U.S. companies wherever they operated both ends of an international route, and foreign companies where they supplied traffic to the United States. Domestic political gains of lower consumer tariffs were more important than the opposition of foreign governments. In particular, South America and liberated Europe were in its sphere of influence so that the FCC saw its actions as legitimate. It interfered in private contracts between companies as it was to do in the 1990s. But the FCC could argue then, as it argued later (see chapter 6) that its actions were not unilateral because they only indirectly affected non-U.S. companies.

Throughout the period, the FCC's regard was inward looking toward Congress and the president, using nationalism, the "free flow of information," and the liberalization of the British Empire as legitimating ideas. The height of the FCC's domestic power came at the Atlantic City conferences of 1947–49. That the U.S. delegation agreed at the Moscow preparatory conference to the binding of all members to the ITU regulations suggests that the FCC saw the mechanism as a means to formalize its power over the companies—and that was its eventual impact. That ITT allied itself with the Pentagon to outflank the FCC, only to be itself outflanked, demonstrates the tensions in the regulatory relationship.

Much FCC energy in the postwar ITU was spent exempting U.S. companies from ITU regulations—thereby expanding its gate-keeping power. But it failed to bring the ITU under the oversight control of the United Nations and (thereby) U.S. control. Having decentralized the commonwealth ownership of communications assets and formal regulatory power over the commonwealth network, the British were able to call on the Dominions as well as European allies to block the U.S. attempt to have the ITU located in New York.

The state-to-state model of the ITU fitted with the growth of nationalism in the postwar period. Although the ITU had accepted end-to-end networking under its "special arrangements" clause, the British attitude toward the operations of U.S. record companies in the United Kingdom was determined by its desire to adhere to the commonwealth resolution of 1942. It both wanted rid of the companies *and* profits from transatlantic traffic. For those reasons it rejected ITT's attempt to use U.S. defense interests to evade the "international formula" and establish a new transatlantic coaxial cable. The British wanted national control, not a "Western Union" model of international communications.

Coaxial cables made the end-to-end U.S. telegraph cables redundant and institutionalized the state-to-state model of the international network. British technological leadership in submarine repeaters meant that the United States could not exclude it from coaxial cables. TAT-1 gave the British its state-to-state model, but the FCC, protecting the international record companies, still refused to allow transmission of telegraph traffic over the cable.

The British-Canadian CANTAT-1 cable was a means of bypassing the FCC. The decline in traffic on the old cables then forced the FCC to agree to the previous British proposal that the record companies first lease lines and then buy into the TAT cables. Even as the United States argued in the ITU that its companies' end-to-end circuits fell under the clause on "special arrangements" exempting them from ITU regulation, the final demise of the end-to-end model was drawing closer with the rise in postwar nationalization of ITT's operating companies. What happened in Chile was evidence of that company's leverage in Washington and part of a tradition of U.S. military backing of its companies' economic interests in South America. In that environment and with ensuing suspicion of foreign investment, it was state-based monopolies, such as that of AT&T, which flourished.

Yet, because of its opposition, ITU telephone regulations did not extend outside Europe until 1958 and, even though signature was compulsory, it was not until 1973 that the United States signed them. From the 1940s on, AT&T, with its view of itself as operating in the public interest, was too powerful for either domestic or international regulation—particularly by an FCC weakened from congressional attacks. Despite its attempt in 1964 to prevent AT&T's monopolizing the provision of international leased lines for both data and voice, the FCC failed to carry through the 1943 act's intention to separate domestic record carriage and voice. As we shall see in the next chapter, AT&T and the FCC became almost synonymous. Meanwhile, despite the FCC's attempt to build an empire under its control, a world telecommunications system built on the sovereignty of nation-states was the lasting legacy of Roosevelt and Churchill's shipboard agreement of 1941 on the self-determination of peoples.

2

Satellites and
U.S. Unilateral Regulation

The introduction of coaxial cables reinforced the existing postwar trend toward nationally based, stated owned monopoly Post and Telecommunications Organizations (PTOs). Whereas the British attempted to renew their previous global hegemony through a round-the-world Commonwealth cable system, satellite technology promised to open up national monopolies to U.S.-owned end-to-end networks. This chapter discusses the introduction of satellite communications in the 1960s and the political process that created first Comsat and then Intelsat—the goal being U.S. control of satellite production and technology.

Comsat, a U.S. company half-owned by the U.S. carriers, controlled Intelsat and was, in turn, controlled by the FCC. The beneficiary was the U.S. aerospace industry. Failing to benefit, AT&T and the European PTOs then colluded to bypass satellites with transatlantic cables. The FCC again became an unacknowledged world regulator. Having originally fought for an Intelsat global monopoly, once the Europeans had developed their own technology and could establish separate systems, U.S. agencies moved to create further competition in the sector. The FCC and the Reagan White House unilaterally expanded domestic regulation of satellite and cable into the international arena. Intelsat's role as global operator and regulator ended and increasing competition led to Intelsat's privatization.

Satellite Communications: The Beginnings of Private Communications Systems

The idea of satellites arose in the 1940s when Arthur C. Clarke, a British Air Force officer, wrote an article in *Wireless World* wherein he discussed the possibility of placing an object in a geosynchronous orbit. Such a system required not just a satellite, but launcher rockets and earth stations, and it was the launcher technology that occupied the U.S. military's Advanced Research Projects Agency (ARPA). During the 1950s, RCA, Lockheed Aircraft Corporation, and AT&T's Bell Labs began satellite research, pitting carriers against aerospace manufacturers. Then the Soviet Union's launch of Sputnik in October 1957 and the failure of a U.S. Vanguard rocket compounded U.S. fears that the Soviets led in the scientific and military race.

In July 1958, Congress passed the National Aeronautics and Space Act, creating the National Aeronautics and Space Administration (NASA) as a research organization and merging space policy with traditional commercial communications. A turf war ensued between NASA and the FCC. By the time the two agencies signed a memorandum of understanding in February 1961, the ITU had allocated frequencies to experimental satellites, and NASA had successfully launched a passive communications satellite—one that could not actively receive, amplify, or transmit radio signals—and had offered launches to private industry. The Eisenhower policy of the private development of communications satellites was well established.

Supporting the traditional divide between private operation and government regulation, both NASA and the FCC attempted to tie the hands of the Kennedy administration, elected with an emphasis on space technology as a symbol of political leadership in the cold war. NASA made policy statements backed by the outgoing administration, and the FCC authorized AT&T's Telstar (a low-earth-orbit satellite) the day before Kennedy's inauguration (Oslund 1977, 163). Then, in March 1961, the FCC's First Report and Order confirmed that satellites as "cables in the sky" should be integrated into the existing structure of the international market.

In April 1961 the Soviet Union's Yuri Gagarin circled the world, and the Kennedy administration suffered the Bay of Pigs disaster. In a public relations response President Kennedy proposed an accelerated space program to land a man on the moon and develop satellites for worldwide communications. On July 24, 1961, he set out U.S. policy. Although the State Department had already agreed on a policy of public ownership, the statement accepted private ownership subject to regulation. The private corporation had to meet policy objectives that included noncommercial global coverage. It had also to allow foreign participation and an ownership structure that would afford maximum competition. For its part, the U.S. government would "examine with other countries the most constructive role for the United Nations, including the ITU, in international space communications."[1] Following this foreign policy agenda, in September 1961 President Kennedy submitted a program for space cooperation to the United Nations and together with the Soviet Union sponsored the General Assembly Resolution 1721, adopted on December 20, 1961, on the Peaceful Uses of Outer Space.

However, NASA and the FCC continued on their domestic policy schedule. Three days after Kennedy's speech, NASA agreed to launch AT&T's Telstar on a cost-reimbursable basis. To prevent an AT&T monopoly, it also contracted with Hughes Aircraft Company for a previously rejected high orbit (20,000 mile high) geosynchronous satellite (Syncom). For its part the FCC, having delayed the public announcement until after Kennedy's statement, dismissed petitions for reconsideration of its First Report and called on the international common carriers to formulate proposals for a joint communication satellite system.

Intended to replicate the status quo, the carriers' plan was for each to carry its investment in satellites in its own rate base and conduct bilateral negotiations on "accounting rates." In addition, so as to ensure that the satellite consortium could not bypass the cable system, each carrier was to solely, or jointly, own ground stations.[2] AT&T was to own as much as 80 percent of the consortium (Kinsley 1976, 4 and 18).

The carriers' FCC-endorsed plan split those in favor and against private ownership of the new satellite corporation. In particular, uneasy at a private company representing the national interest, the State Department argued for a government-owned corporation, but "its views were pressed with only a modicum of skill or energy" (Chayes 1971, 44). The Kennedy bill proposed instead that the federal government supervise through the State Department the establishment of a single privately owned corporation—Comsat—in which no one person or firm could own more than 15 or 20 percent (Kinsley 1976, 5). But the administration was forced to compromise with Senator Robert Kerr of Oklahoma, the spokesman for the communications carriers. The Kennedy bill gradually came to resemble the Kerr bill.

Meanwhile, on July 10, 1962, NASA launched AT&T's Telstar satellite, which demonstrated the first live transatlantic television broadcasts. To NASA's distaste, AT&T itself had negotiated bilaterally with European PTOs and the Japanese to provide earth stations. AT&T was set for a complete monopoly of international communications.

The Comsat Bill was passed in haste engendered by Soviet competition and also the failure of the Kennedy legislative program. As Abram Chayes (1971) comments,

> To be sure, the legislation makes appropriate genuflection to "world peace and understanding," to "providing services to economically less developed countries" and to "non-discriminatory access to the system." But, although President Kennedy's statement had spoken of "the U.S. *portion* of the system" most Congressmen, I think it fair to say, conceived of the proposed global system as a United States show. It was to be not unlike the Union Pacific, another company chartered by Act of Congress: anyone could ride, but it was our railroad. (44, emphasis in original)

The bill came under sustained attack in the Senate. For the last time in fifty years a small number of legislators, led by Senator Estes Kefauver of Tennessee, Senator Wayne Morse of Oregon, and Congressman Emannuel Celler of New York, with support from the American Communication Association, Assistant Attorney Lee Loevinger, and academics Dallas Smythe and Herbert Schiller, fought for the concept of public ownership along the lines of the Tennessee Valley Authority (Galloway 1972, 52–57).[3] They not only objected in principle to private ownership but also argued that in practice the FCC was unable to regulate AT&T.[4] But AT&T had swung behind the Kennedy Bill and opponents were heavily outnumbered. To end the filibuster on August 14, 1962, Congress voted 63-27 for a cloture on debate for only the fifth time in its history. The bill was passed with only eleven senators voting against.

The Communications Satellite Act of 1962 established a private corporation with 45 percent of its stock reserved for a consortium of the major international carriers—AT&T, ITT, General Telephone and Electronics Corporation (GTE), and RCA Communications Inc. Aerospace companies and Western Union were excluded from the consortium. The remaining 50 percent of the capital was to be offered to the public with foreign share ownership limited to 20 percent and that of aerospace firms limited to 10 percent (Kinsley 1976, 6). The new Satellite Corporation was charged with creating a commercial satellite system as soon as possible—then expected to take around four years—but some public service goals remained.

Section 102(a) of the legislation stated that Comsat was to serve the communications needs of the United States and other countries, but that "care and attention will be directed to economically less developed countries and areas."[5] Yet there was no clear understanding of how international cooperation could be achieved. A compromise in the final bill reduced both the power of the State Department and of the president over the foreign relations of the corporation. Lack of a clear division of authority between Comsat and the State Department was to lead to another turf war.

Placing Comsat under the 1934 Communications Act, the final act made the FCC responsible for the supervision of the system's technical development and for the capitalization of Comsat. It was to "prescribe such accounting regulations and systems and engage in such ratemaking procedures as will insure that any economies made possible by a communications satellite system are appropriately reflected in rates for public communications services" (cited in *Harvard Law Review* 1962, 390). Comsat therefore came firmly under the regulatory objectives of the 1934 act that prioritized the U.S. consumer.

Comsat—A Private, Unilateral Operator and Regulator

From a European perspective the hurried Comsat legislation made previous plans for a Commonwealth/European satellite redundant. British and Canadian officials met U.S. officials during the passage of the act to discuss cooperation in a commercial satellite system.[6] Canadians originally favored executive control by either the United Nations or the ITU under the U.N.'s resolution of December 1961—a proposal with which the British disagreed.[7] The U.S. side was also against involving the United Nations or the ITU. The British and Canadians then argued for an international system in which national PTOs would hold shares in proportion to intended use and "those not wishing to have ownership should have access to the system."[8] The model was an extension of the bilateral TAT agreements to the multilateral. But because the State Department was not prepared to put the discussions on the record, the British doubted whether the United States really intended international cooperation.[9]

Similarly, the State Department was suspicious of British and Canadian intentions.[10] The British Commonwealth had rebuffed AT&T's request to participate in its coaxial cable between Canada and Australia. Chayes would later say: "British interests and other European interests had thought . . . they could get in with . . . another generation of cables and put this whole thing off" (cited in Schiller 1969, 137). In December 1962, rather than allow the British to report back on the talks to members of the Conference Européen de Poste et Telecommunication (CEPT), the State Department spoke to CEPT directly.[11] But then Comsat's first chairman, Philip L. Graham, the publisher of the *Washington Post,* "strenuously objected to what he saw as the interference of the State Department in the international aspects of Comsat's plans" (Galloway 1972, 81). In particular, he objected to the State Department briefing the Europeans.

Comsat officials, using AT&T's terminology of a "cable in the sky," first attempted to negotiate separately with London, Paris, and Bonn. Comsat offered a joint venture agreement or the lease of channels in an American-owned satellite. Although in agreement with European proposals for an international institution, the State Department could do little. It "had to operate in terms of a statute and Congressional background that seemed to contemplate United States hegemony, at the least, in the global system" (Chayes 1971, 45).

In response, pushed by the British, CEPT set up the European Conference on Satellite Communications (CETS) to coordinate European policy.[12] France had already stated that it intended to develop its own launcher. France, West Germany, and Britain all wanted to ensure that in return for cooperation with the United States, they would receive equipment contracts and transfer of technology.

In February 1963, Leo D. Welch, formerly the chairman of the board of Standard Oil, became the new chair and chief executive officer of Comsat. Like his predecessor, he wanted the company to own both the satellite system and ground stations and to negotiate bilaterally (Galloway 1972, 83–85). Whereas the State Department was trying to convince other countries that Comsat was not a U.S. monopoly, Welch wanted a monopoly end-to-end system run from the United States.

At the same time, Joseph V. Charyk became president and chief operating officer of Comsat. He had been in charge of the Air Force reconnaissance satellite program in the Eisenhower administration before heading the Kennedy Administration's classified National Reconnaissance Program, covering all overt and covert satellite surveillance. Charyk therefore linked Comsat with the Pentagon and intelligence services.

Differences between the Department of State and Comsat led in June 1963 to an Ad Hoc Communications Satellite Group to work with Comsat in negotiating agreements for a global satellite system (Oslund 1977, 168). Chayes says,

> The Department's approach was . . . to contrive repeatedly to expose the officers of the company to situations, meetings and conferences where they could experience, as uncomfortably as could be arranged, the international realities of the situation. Gradually this process took its toll, and Comsat was brought to the realization that some form of international mechanism for ownership of the international system was essential. It also brought the United States government, in return, to support Comsat's demands for dominating authority in the international consortium. (1971, 45–46)

Following the failure in February 1963 of the first Hughes geosynchronous satellite, Syncom II began regular service on August 16, 1963, and was able to transmit voice, teletype, facsimile, and data. NASA became committed to geosynchronous, rather than low-earth orbit, satellites and persuaded Comsat to launch a version of Syncom using commercial frequencies (Whalen n.d., 27). AT&T's role as a low earth orbit satellite manufacturer and operator became sidelined. It once more became a champion of cables.

From the European side, in July 1963 Britain and France, putting aside their differences over French proposals for a regional satellite system, made a joint approach to the United States. They proposed that satellite communications should be organized on an international basis in such a way that the European countries could participate in the design of the system, share in its ownership, play a full part in its management, and have the opportunity to provide equipment.[13] In reply the State Department indicated support for these principles, but not exclusive to Europe.[14]

Eventually, in November 1963, Comsat approached CETS with proposals that would permit foreign shares in the ownership of a satellite system and allow foreign owners to participate in management of a system under the day-to-day management of Comsat.[15] President Kennedy backed Comsat in these negotiations. On November 20, 1963, he invited foreign countries to join in establishing a global satellite system with Comsat in charge of the U.S. end. Addressing his speech to ongoing Brussels negotiations between CETS and Comsat, he confirmed "that all countries which wish to participate in the ownership, management and use of this system will have an opportunity to do so" (cited in Pastore 1964, 122). Kennedy was assassinated the following day.

The Creation of Intelsat—Interim Arrangements

Negotiations with the Europeans eventually established the interim arrangements for a Global Commercial Communications Satellite System (Intelsat) in 1964. The British Post Office was won over in December 1963 by an AT&T letter to Comsat "announcing that it would rely on satellites to meet its new trans-Atlantic communications needs for at least the rest of the decade."[16] The British Labour government's view became "that the only way of preventing an American monopoly in this sphere is to join a partnership with the United States and other countries and so secure the right to influence the course of events."[17]

However, the Europeans had not given up thoughts of their own system. During the negotiations they proposed the inclusion in the agreements of

the phrase "nothing should prevent any party from creating additional satellite systems if required to meet unique government needs or if otherwise
required in the national interest," which copied part of Section 102(d) of the
1962 U.S. act. But the U.S. negotiators refused the clause as undermining
the U.S. monopoly, and the Europeans held no bargaining power (Naraine
1986, 264).

The interim arrangements, which related to the "space segment" of the
satellite system, consisted of an intergovernmental agreement and a "special
agreement" between telecommunications entities.[18] The interim organization
had no assembly of members, but was headed by a twelve-member Interim
Communications Satellite Committee (ICSC) on which voting power was
directly related to investment. The model was similar to that of the World
Bank. Investment quotas were based on current shares of international traffic. However, the inclusion of domestic traffic over two thousand kilometers
from the U.S. mainland to Guam and Puerto Rico inflated the U.S. quota.
New signatories were limited to 17 percent, ensuring that the U.S. share could
not fall below 50.5 percent. Voting provisions gave the United States a perpetual veto.[19] Although in 1964 the Soviet Union held discussions with the
United States, it was mainly interested in television, and low international
telephone traffic flows would have given it little influence under the interim
arrangements' formula.[20]

The arrangements accepted Comsat's position as the designated manager
of the system with the power to prepare budgets and negotiate contracts. Yet,
even after their agreement, Comsat and the Pentagon lobbied to alter the
interim arrangements.[21] The Pentagon proposed that it should share a satellite
using separate frequencies to those used for civilian purposes as a "special"
user outside the regulatory authority of Intelsat.[22] This suggested organization
was similar to the ITU agreement, in which "special arrangements" were outside ITU regulation. Comsat also wanted the interim arrangements revised
to strengthen its monopoly over technology, budgets, and procurement. But
both the State Department, which had never been enthusiastic about military
involvement, and the FCC opposed the last-minute revisions.

Nevertheless, not wishing to lose the $35 million Pentagon contract, Charyk
attempted to use Congress to lobby other Interim Committee members and
then attempted to bypass the voting procedures of the interim arrangements
(U.S. Congress, House of Representatives 1965, 103). Finally Secretary of Defense Robert McNamara announced in July 1964 that the U.S. military would
proceed to develop its own communications satellite—the beginning of an
entirely separate military system of satellites.[23]

Problems with the Interim Arrangements

Early Bird, launched into geostationary orbit in 1965 and able to carry 240 telephone conversations or one television channel at a time, ensured that the ensuing 15 percent annual growth in traffic could be accommodated. Without it the transatlantic cables would have been full by 1966 (HMSO 1964, 5). Net operating income allowed a rate of return on investment of 5 percent over the three years to 1969 and of 12 percent in 1969. For its investors Intelsat was a commercial success.

But in other ways the interim arrangements were unsuccessful. Intelsat was located in a small portion of the floor housing Comsat and because Intelsat had no juridical personality, Comsat *was* Intelsat. In turn, because Comsat was Intelsat and Comsat was a domestic company, the FCC regulated Intelsat's contracts. Whereas the interim arrangements stated that when comparable tenders were made, contracts should be distributed in approximate proportion to the respective quotas of the signatories, the FCC demanded competitive tendering on price.[24] The process was seen as acting against European contractors.[25]

Relations on the Interim Committee were also fraught. The French complained about the British Post Office representatives' uncritical support for Comsat.[26] Then, Comsat's attempts to appoint a company employee as secretary of the Interim Committee unified the Europeans into a joint démarche to the U.S. State Department. The Foreign Office commented, "The general feeling in Europe particularly . . . is that the American COMSAT Corporation . . . is abusing its dominant position and denying the cooperative, international character of the ICSC."[27] In addition, in September 1965 President Lyndon Johnson approved NSAM338, preventing U.S. assistance for foreign satellite systems competitive to Intelsat—except in the case of security needs. Despite European frustration, the United States was intent on a commercial monopoly (Sebesta n.d., 3).

The interim agreements had compromised between European desire for a full-blown international organization and U.S. reluctance. Definitive Arrangements were scheduled to enter into force by January 1, 1971, and discussions therefore began in 1969 on a permanent organization. The Europeans wanted a reduction in U.S. voting power, more equipment contracts, and the replacement of Comsat as manager.[28]

Under the interim arrangements, although thirty-nine more countries had joined the original eighteen members of Intelsat, Comsat retained its veto. In addition, there was a conflict of interest between Comsat's responsibilities to the U.S. government, to its shareholders, and to Intelsat.[29] Yet Comsat, backed by the U.S. government, was determined to maintain its role as manager.[30] Following the passage of the U.N. Treaty on Outer Space, in mid-1967 President Lyndon Johnson intervened. He accepted that there could be ceilings on voting power, a formal assembly, internationalization of Comsat personnel, and a wider distribution of contracts, but he argued that "the size of our investment has made it logical that Comsat serve as a consortium manager."[31] The U.S. State department's protection of Comsat's dual role prevented discussion of British proposals for an intergovernmental agreement.[32]

Meanwhile, the Europeans were split again.[33] A new French government wanted satellites owned and operated by separate consortia and Intelsat to be a regulatory body—a proposal rejected by the United States, which wanted Intelsat to both operate and regulate.[34] The threat of separate systems and demand for U.S. domestic satellites brought the United States to accept a CETS proposal in 1968 that "groups of countries in a closely integrated land mass like Europe should have corresponding rights to provide a regional satellite carrying international traffic, to the rights which a single large country like the U.S.A. might have to provide a satellite for domestic traffic."[35] This was the first move toward fragmentation of Intelsat's monopoly. Then in August 1968 the Soviet Union announced at the U.N. Conference on the Exploration and Peaceful Uses of Outer Space that it would set up an Intersputnik system based on equal voting rights. There was concern that without an intergovernmental agreement giving them more influence, developing countries might pull out of Intelsat.[36]

The Creation of the International Telecommunications Satellite Consortium

The plenipotentiary to settle the permanent organization—the "Definitive Arrangements" of Intelsat—was held in Washington in February 1969 but adjourned—first until 1970 and then until 1971. From 1968 negotiations were handled by an Intersessional Working Group consisting of twenty-five to thirty delegates. The working group split into two alliances—the 54 Group (led by Mexico and backed by CETS) and the 45 Group (led by the United States closely aligned with Japan and Australia). Commonwealth countries were split between the two, with India, Canada, and Caribbean and African countries working with the United Kingdom. The British, then applying for entry to the European Union, aligned with the Europeans.[37]

Because the definitive arrangements would weaken its role, Comsat campaigned to retain the interim arrangements.[38] The problems of a private corporation heading international negotiations were underlined in February 1970 when hardliners on the Comsat board sacked the chairman, who, as the U.S. representative, had accepted a "package" of proposals with which the board disagreed.[39] Comsat's behavior caused British-U.S. bilateral talks at the Secretary of State level and a joint démarche by the French, German, British, and Swiss governments. They protested that the phasing out of Comsat, while "a domestic and commercial problem peculiar to the U.S.A.," should not be "an insuperable obstacle in negotiating an international agreement with some 70 other governments."[40] At one point the head of the British delegation "found it necessary to speak to the leader of the American delegation, plainly and on the record, though in private, of British disquiet over American tactics." Ambassador Abbott Washburn then assured the British "that the United States genuinely wanted Definitive Arrangements." He "offered some hope of bringing Comsat under control."[41]

An independent management review undermined Comsat's position. It concluded that Comsat had prevented the Interim Committee from considering long-term policy objectives, and its financial procedures had been inadequate to provide for "meaningful control and evaluation by the I.S.C."[42] At the end of 1970 the head of the British delegation reported that Ambassador Washburn had agreed that "Comsat, like Stalin in Europe in 1945, had succeeded in destroying by their behaviour so much of the good will . . . which they could have capitalized upon in the definitive arrangements."[43]

As the negotiations neared their end, conflict arose over the powers of the intergovernmental Assembly of Parties. The U.S. government and Comsat wanted to circumscribe the assembly's powers, whereas the French felt that its powers were too weak.[44] The British, although irritated by the last-minute French intervention, hoped that U.S. isolation on the issue would force the Department of State to control Comsat.[45]

The U.S. government backed Comsat in objecting to regional satellite systems. The original European space plan had been to launch its own satellites with European Launcher Development Organisation (ELDO) rockets. France and Germany had hoped to improve the European negotiating position with the development of their own Symphonie satellites. But when the European launcher failed during the 1960s, U.S. bargaining power increased. In 1968 NASA had refused to launch the Franco/German satellites unless they were "experimental"—in other words, no threat to Intelsat. The final agreement, which allowed separate systems only if they did not cause "significant economic harm" to Intelsat, could be viewed either as a partial victory for the United States as a result of NASA's implicit threat (Snow 1976, 102) or as a partial victory for the Europeans brought about by Japanese influence on the United States.[46] But in 1971 the Europeans were to find that U.S. interpretation of the voting provisions hardened under industry and Comsat pressure. In order to escape a U.S. veto they would need a two-thirds majority of the assembly in favor of a regional satellite (Sebesta n.d., 10–11).

Despite opposition to internationalizing Intelsat's management and to forfeiting its veto, the United States eventually conceded on both.[47] Its isolation also led it to agree that while the one nation–one vote assembly would be mainly recommendatory and advisory, it would have power to decide on proposed amendments to the Intergovernmental Agreement and on proposed specialized services. France withdrew from the final agreement, formally on the grounds of weak assembly powers, but probably because of its wider space policy concerns.

The new agreement signed on August 20, 1971, came into force on February 12, 1973. Intelsat's function remained satellite transmission of point-to-point public telecommunications services and certain defined domestic services of an "international" character. The board of governors still held "the responsibility for the design, development, construction, establishment, operation and maintenance of the Intelsat space segment" (Article X), but the U.S.'s voting power was capped at 40 percent—and it no longer had a veto (Article IXV). In addition, under the new system, investment shares were to be regularly adjusted to reflect current traffic (Intelsat 1996).

Over a period of six years an international executive organ headed by a director general responsible to the board of governors was to replace Comsat as manager of the system. During those six years Comsat would also be put under contract to Intelsat and monitored by the director general. From the U.S. perspective, Abram Chayes described the definitive arrangements as "the worst of both worlds." The United States had lost its voting majority and veto, and Comsat had lost its position as manager (Chayes 1971, 47).

European manufacturers had been disappointed with the procurement system under the interim arrangements. Although overall European participation as subcontractors had risen from virtually nothing on Intelsat III to 26 percent for the first four satellites of the Intelsat IV series, most Intelsat procurement went to U.S. companies (U.S. House of Representatives 1965, 100). *Europspace,* formed in 1960 from about one hundred European industrial companies, had lobbied for the allocation of procurement according to national quotas of capital. A rewording of the clause on procurement under the definitive agreements so that, where two or more tenders were equal, the contract would be awarded "so as to stimulate . . . international competition," led the Europeans to hope for more orders.[48] But then in 1970 Comsat declared that non-U.S. participation in Intelsat IV "had resulted in at least a 5 percent increase in overall costs" (Snow 1976, 125). Europeans gained only 10 percent of contracts in the later four satellites. Hughes Aircraft, Ford Aerospace and Communications Corporation, and TRW Inc. were the sole prime contractors between 1963 and 1986 (Pelton 1984a, 78).

In effect, Comsat and Intelsat were equivalent to a PTO, using procurement to benefit domestic (U.S.) manufacturing industry and passing the resulting costs on through averaged tariffs to the customer. In this case the beneficiaries were primarily the U.S. aerospace companies and the losers were those customers—developing countries—most reliant on Intelsat. Small wonder, then, that developing countries wanted training programs and scholarships to compensate for their contribution to the research and development of U.S. industry and were reluctant to pay for a European technological learning curve (Kildow 1973, 53). But the FCC's parallel policy on submarine cables also added to developing countries' costs.

The FCC: Regulator of International Satellite Communications

Although Comsat/Intelsat was inaugurated as both an operator and regulator of international satellites, it was the FCC that controlled submarine cable across the Atlantic and Pacific. Under a 1954 executive order, the commission had only to take advice from the Department of State on cable landing license approvals.[49] Because Intelsat concentrated on point-to-point transatlantic communications, until 1973 (when the definitive arrangements came into effect) the FCC directly regulated Intelsat through Comsat, and indirectly regulated it through licenses for transatlantic cables.

The U.S. carrier companies' view was that they had not invested in Comsat for it to compete with them. When, in 1964, the FCC awarded the first three earth stations and their local loops to Comsat, it faced such carrier opposition that it eventually agreed shared ownership between Comsat and the carriers. The decision allowed the carriers to count the earth stations in their rate base (on which a regulated company's tariffs depended), to take 75 percent of any return on them and to maximize the local loops between them and the local exchange. It also meant that Intelsat became dependent on AT&T (Murphy 1971, 412).

But FCC limitations on satellite competition went further. Under the 1934 Communications Act, AT&T could make a 12 percent rate of return on capital before it had to reduce tariffs to domestic customers. AT&T could lower its rate of return by investing in capital projects, such as international cables, which indirectly kept consumer tariffs high. For instance, the first four transatlantic submarine cables with 240 voice-grade circuits cost $133 million; TAT-5, with 720 circuits, cost another $80 million. Each cable was then depreciated over twenty-four years. TAT-6, laid in 1976, carried 4,000 circuits at an annual revenue requirement of only $8,600, but, when averaged with the historic cost of the five previous cables, the annual revenue requirement became $16,600 per half circuit. AT&T then set tariffs to cover this requirement (Stanley 1977, 80).

In contrast satellite depreciation took place over seven to ten years, and cost advantages lay in their multiple- rather than single-route capabilities. As Walter Hinchman (1971, 28) explains, the cost of a link between two earth stations is independent of distance, and the cost per route decreases as the number of routes per station increases. Hence "per-circuit costs of satellite interconnection within a given coverage zone are . . . inversely proportional to the total traffic volume over all routes served." A reduction in total volume of traffic resulted in an increase in the average cost per route. And since the largest proportion of world point-to-point traffic in the 1960s was that across the Atlantic, any diminution in Intelsat's carriage of that traffic inevitably affected its overall averaged tariffs.

For the FCC, satellite entry into the international market came at a time when the record carriers were already threatened from AT&T's coaxial cable. From Intelsat's perspective, the 1964 FCC decision to reserve alternate voice/data services over leased lines to the international record carriers and to allow the record carriers to invest in TAT-4 had two effects: from 1964 the cable lobby included all the international carriers; and AT&T, now limited to international telephone, was under increased pressure to fill its cables.

Initially the capacity on Intelsat's satellites was much greater than that possible over submarine cables. Intelsat I and II launched in 1965 and 1967, respectively, carried 240 voice circuits or one TV circuit each. Inevitably, Comsat looked to large users to lease capacity, particularly the U.S. Department of Defense. But, despite provisions of the 1962 act that allowed Comsat to serve "authorized users and the U.S. government," complaints from the record carriers led the FCC to ban Comsat from serving any customer directly (Oslund 1977, 153).

Subsequently, Comsat was also refused permission to deal directly with the television networks.[50] Despite feelers from Congress and the Johnson White House, Comsat never appealed the decision—perhaps related to the carriers' representatives on its board. The "authorized users" rule made Comsat/Intelsat dependent on factors outside its control—demand generated by the carriers and competition from submarine cable. In 1964, AT&T, owning 58 percent of those Comsat shares reserved for carriers, had declared that if satellite circuits were provided at a cost similar to cable, it would prefer satellites over cables "until North Atlantic routes were served by approximately equal numbers of cable system voice circuits and satellite system voice circuits."[51] The assumption was therefore that half the transatlantic traffic would go by satellite. But in March 1966, as AT&T filled up its new TAT-4 cable, the FCC asked the company to lower its international rates so as to stimulate demand for unfilled capacity on Intelsat I.[52]

Nevertheless, the FCC continued to approve new cables. In 1967 AT&T applied for a 720–circuit TAT-5 cable to Southern Europe. It argued that because of the arrogant behavior of Comsat, the European PTOs perceived satellites as representative of U.S. domination and therefore preferred cable (Kinsley 1976, 69–71). The FCC, with Commissioner Nicholas Johnson dissenting, overruled Comsat's objections that Intelsat III, to be launched in 1968, would carry 1,500 additional voice circuits, and that TAT-5 would add $250 million to customer tariffs over twenty-five years. Instead, it ordered the carriers to reduce their tariffs by 25 percent on the increased rate base.

The FCC also obliged the carriers to "fill satellite and cable facilities at the same proportionate rate so that 100 per cent utilization of these facilities would be reached at the same time."[53] But this order did not mean that where one thousand circuits were provided by satellite and one hundred by cable, carriers had to use ten satellite circuits to one cable circuit. Once TAT-5 was built, the record carriers elicited twenty objections to the proportionate fill requirements from the Europeans, refused to use TAT-5 until the dispute was resolved, and refused to lower rates. Led by ITT, they achieved a postponement of the rate reduction.

In 1971, at a time when ITT was contributing illegally to Republican Party coffers, the company went directly to the Nixon White House, one week before the launch of Intelsat IV, to demand its delay. Unsuccessful at the White House, ITT then successfully demanded that the FCC exclude Intelsat IV from proportionate fill. When AT&T then negotiated a ratio of five satellite circuits to one cable for Intelsat IV, the record carriers were successful in pressing the Europeans to object to this ratio. Once again the FCC backtracked so that proportionate fill became one to one and the record carriers were allowed to fill their quota on TAT-5 before filling Intelsat IV (Kinsley 1976, 93). In sum, the FCC favored the carriers and protected their profits against Intelsat. The beneficiaries were European PTOs whose transatlantic cables were predominantly paid for by AT&T and U.S. customers.

The effect of these various carrier maneuvers was that most of Comsat's circuits were unused. In 1970, of 2,129 circuits that Comsat had leased on two Intelsat III satellites, 1,740 (80 percent) were empty. In contrast, of the 488 circuits AT&T controlled on TAT-5, 49 percent were empty. To use its spare capacity in 1974, Intelsat began leasing to developing countries for domestic service. By 1979, 16 countries had taken five-year leases at a cost of $1 million per year (*Satellite Communications* 1979, 9). Intelsat then announced greater flexibility—short-term leases of three months for domestic use only, extendable on a month-to-month basis. Meanwhile, the Nixon White House had intervened in the domestic satellite market.

Domsat: A Unilateral U.S. Decision

The U.S. domestic satellite issue had been around since 1966 when the Ford Foundation first proposed a satellite to distribute television programs and fund educational television. Much of the early FCC debate was focused on Comsat and AT&T's role, and it delayed decision for fear of its impact on Intelsat negotiations and a European separate system. By 1970, it became evident that a European system was inevitable. A Nixon administration group, headed by Clay T. Whitehead, later to head the White House Office of Telecommunications Policy, issued a recommendation to the new FCC Chairman, Dean Burch. "[A]ny financially qualified public or private entity, including Government operations, should be permitted to establish and operate domestic satellite facilities for its own needs . . . or to be used in providing specialized common carrier service on a competitive basis."[54] In the ensuing press conference it was made clear that "specialized" services, defined as television or high-speed data transmission, were to be unregulated. Dan Schiller (1982, 91) argues that it was large business users, formed into "potent, politically directed groups" through the Corporate Committee of Telecommunications Users and the International Communications Association, who demanded that computer transmission remain under their control. FCC regulation threatened their existing investment. Here, then, was the beginning of the "basic" and "enhanced" division of the FCC's 1980 Computer II decision (see p. 83).

The Whitehead memo was unilateral in application. Although it recognized that other Western countries might have an interest and that "a question of United States monopolization could conceivably arise," it minimized the possibility.[55] However, according to Chayes (1971, 49–50), "There was no effort to consult with affected countries before . . . the Memorandum was issued." Chayes argued that the United States should "stop trying to have it both ways—to acknowledge the essentially international interests in satellite communications and at the same time to preserve substantially unfettered national freedom of action." But, still struggling in the Intelsat negotiations, the reaction of other countries was muted.

With the exception that AT&T was barred from using satellites for its public switched network for three years, the FCC eventually adopted the Whitehead "open skies" policy. Western Union was the first to launch, with Westar in 1974, but data communications via satellite were slow to expand. Instead, first cable television and then Home Box Office pay television filled the unused domestic transponders.

IBM was able to enter the data transmission business when, in 1980, with Comsat and Aetna Life Insurance it created Satellite Business Systems (SBS). In turn, SBS, using the higher Ku-band that required smaller earth stations, led to the Vsat model adopted in the mid-1980s by Intelsat to provide company-to-company leased lines (i.e., private networks). From 1981 Cable & Wireless also provided end-to-end data communications service via Intelsat and its U.S. and U.K. subsidiaries. The liberalization of the U.S. domestic satellite market then led to demands for a similar liberalization of international satellite communications and to commercial satellite television in Europe.

AT&T Rules: North Atlantic Facilities Planning

As the Intelsat discussions were continuing, AT&T announced in 1970 that it would build ten more transoceanic cables before the end of 1979. It applied for permission to lay a 750–circuit TAT-6 to be shared with the French and German PTOs. But of the total $86 million, $70 million would go on AT&T's rate base. Comsat costed each circuit at $421,696, compared to the $1,450 of an existing Intelsat circuit (Kinsley 1976, 98). Whitehead's Office of Telecommunications Policy intervened. It argued that without new cables, satellite rates could be substantially reduced, and it published Hinchman's (1971) study on the cost advantages of satellites. The FCC refused the TAT-6 application.

The French reacted furiously to the veto, saying that the FCC intended "to favor excessively a system it dominates—Intelsat."[56] A series of meetings with

CEPT and Canada began in September 1971, and the FCC issued a statement of policy on planning in the North Atlantic Area. Its objectives included the continued development of satellite and cable; regard for efficiency, economy, diversity, and redundancy; and adherence to "international comity with the consequent minimization of artificial cable/satellite formulas."[57] But in November 1972, following ITT's opposition to the 13,000–circuit Intelsat IVA generation of satellites, the FCC ordered Comsat (then still the manager of Intelsat) not to ask for approval from the Intelsat board. However, the FCC had no direct power to prevent Comsat from doing so, and ITT had already sold its shares in Comsat. Comsat disobeyed (Kinsley 1976, 93).

Meanwhile, FCC allocation formulas for North Atlantic cable and satellite traffic led to the institutionalization in 1974 of the North Atlantic Consultative Process. Intelsat was present at the first of these meetings, invited by the host country, Germany, but it was to receive no further invitations (Pelton 1977, 109). When the record carriers were excluded from the June 1975 meeting, ITT filed a complaint under the 1976 Government in the Sunshine Act (one of a number of laws intended to create openness in government), winning the case in 1983.[58] Henry Geller, the Carter administration's communications specialist, was later to argue that this case demonstrated how the FCC's domestic processes made it unsuitable to undertake international negotiations (Geller 1984, 67).

Following the European issue, a U.S. Office of Telecommunications Policy report proposed in 1975 that the FCC "drop its allocation procedures for satellite/cable traffic . . . and allow satellite utilization charges to be capitalized and thus included in the rate base of international carriers."[59] In other words, Whitehead wanted to provide incentives to Intelsat's use. But the FCC opposed these proposals and on February 19, 1976, adopted guidelines in which it said that the "TAT-6 cable should not be unduly loaded in its early years while readily useable capacity remains in existing Atlantic Satellites having a much shorter lifetime than the TAT-6 cable."[60]

AT&T and the U.K. Post Office were very critical of this report. They both wanted more traffic to go via cable and filed a statement with the FCC alleging that "international principles of comity were being violated."[61] At the subsequent North Atlantic Planning Meeting in March 1976, the U.K. Post Office (as the European representative) said it would stop all cable growth until accommodation on the use of satellite and cable circuits was reached with the United States (Pelton 1977, 110).

In reply, the FCC stated that FCC policy was not to favor one technology over another "but to achieve the most efficient utilization at the lowest cost."

And it would continue to evaluate all authorizations for new major facilities "in the context of a comprehensive long range plan" (cited in Yurow 1983, 114). In fact, it continued to make unilateral decisions outside the planning meetings.

Then, with the proposed introduction on the North Atlantic route in 1979 of TAT-7 with 4,000 circuits and a 12,000–circuit Intelsat V in 1980, fears of overcapacity renewed demands for an integrated system of planning. But potential coordination in North Atlantic Planning Meetings succumbed to domestic politics. In 1981 the FCC decided that no new cable systems would be required until 1988, when a fiber-optic cable should be laid and the Intelsat VI series introduced. The two together would produce 160,000 voice-grade circuits, compared to a projected requirement of 37,161 circuits in 1990 (Naraine 1986, 294).

Such overcapacity adversely affected developing-country users of Intelsat, but they had no representation on the matter. Although the cost of a full-time telephone circuit from Intelsat had fallen from $60,000 per year in 1966 to $10,000 by 1981, with full capacity the cost could have dropped further. But from the point of view of the Europeans in particular, it was in their interests to keep Intelsat tariffs high, gain a 14 percent return on capital, and benefit from AT&T subsidized cables.

Although it was true in 1984 that for those twenty-four developing countries leasing transponders for domestic service Intelsat provided a lifeline, not all developing countries believed that Intelsat had done all it could for them. Kinsley (1976, 115) suggests that carrier members on Comsat's board initially refused it permission to give technical and financial assistance to the construction of earth stations in developing countries. According to Joseph Pelton (1984b, 416), planning low-cost communications in developing countries was hampered by the split between the ITU, UNESCO, the World Bank, and Intelsat, and between mass media and telecommunications experts. While true, it was the FCC as the virtually unrecognized regulator of Intelsat that had the greatest impact.[62]

U.S. Domestic Policy: Unilateral Application

President Carter's election in 1977 coincided with a general trend toward domestic competition as a central plank to fight oil-induced inflation. The Carter administration called for "legislation to encourage competition by deregulating fully competitive markets . . . such as barriers imposed to prevent international carriers from competing in national markets." It argued that changes in technology "run ahead of the regulatory scheme."[63]

In 1976 the FCC declared carrier-imposed restrictions on domestic "resale" unlawful. Then, in December 1979, having learned that the international record carriers had been earning pretax profits of between 35 percent and 58 percent, the FCC decided to introduce competition in the sector. In response to complaints from the record carriers that Consortium Communications Inc. and International Relay Inc. were offering international services over leased telegraph lines, Charles Ferris, the FCC chair, suggested that facilities ownership, together with "restrictions on types of third party uses of international lines, and the continuation of the distinction between record and voice services," might all need reevaluation.[64] The FCC then determined that data communications provided by these carriers should be allowed in the interests of competition.

In 1980 in its Computer II decision the FCC first introduced the conceptual division between domestic "basic"—the "common carrier offering of transmission capacity for the movement of information"—and "enhanced" services involving the processing or alteration of the message by computer (Zarkin 2003, 284). Under this political construction of markets, "basic" services of voice telephony were considered to be monopolies at the local level and therefore subject to regulation, whereas "enhanced" data processing services were considered to be a competitive market and therefore not subject to regulation.

By replacing the dichotomy of "voice" and "record" and that of "domestic" and "international" with the concepts of "basic" and "enhanced" services applicable to both the national and international arenas, the new division implied that companies selling "enhanced" services could sell them in the international as well as national market. GTE Telenet and Tymnet were then licensed to provide data communications to the United Kingdom and Western Europe (U.S. Congress, House of Representatives 1982, 246 and 290).

In the next three years the FCC ruled that it was no longer bound by section 222 of the Communications Act of 1934 to distinguish between domestic and international carriers. It allowed the record carriers to expand their operations to twenty-one additional domestic gateways; allowed Western Union International Inc. (Xerox) into the international market; and under the Record Carriers Competition Act of 1981 set out to develop a fully competitive domestic and international market. It also removed the divisions between international voice and nonvoice services; allowed Comsat to compete with the international carriers for overseas traffic, and required all international telex systems to be interconnected to each other and domestic carriers (Yurow 1983, 107 and 114–16). By 1982 all FCC regulatory support for a state-to-state system had ended.

A further FCC decision in 1980 allowed "resale" in the international market. But international "resale" over leased lines went against the state-to-state–based structure of the ITU (Eward 1985, 23). Leon Burtz, the director of the CCITT, wrote to FCC Chair Mark Fowler to point out that the United States was a party to the ITU agreement. AT&T took fright at the prospect of European reprisals and the Department of State found itself called on to control the unilateral tendencies of the FCC. In October 1980 the FCC reassured Study Group III of the CCITT that in "no way does the United States through the FCC wish to take unilateral action that would harm the exceptional international cooperative effort that characterizes the CCITT deliberations."[65]

But the die had been cast. U.S. policy now turned to evading and then breaking the rules of the ITU (see chapter 3). First, however, the U.S. satellite industry focused on breaking the rules of Intelsat.

Changes in the Market

Despite FCC regulatory decisions in favor of cable, by 1985 Intelsat carried two out of every three intercontinental telephone calls. With 108 members, its satellite system provided 65,000 voice circuits and virtually all regional and international television relays. One third of the ownership and use of the Intelsat system was by developing countries (Pelton 1984a, 77–78).

But U.S. compromise with European demands had allowed Article XIV(d) of the Definitive Arrangements to legitimize competition, provided it did no "significant economic harm" to Intelsat. The interpretation of this clause depended on the scope of existing services. Comsat/Intelsat's concentration on the point-to-point transatlantic voice market during the 1960s and 1970s provided an opening for domestic and regional satellites and for satellites focused on television transmission.

As well, the United States overplayed its hand in the launcher market, a factor precipitating the 1975 creation of the European Space Agency and the development of the Ariane launcher (Sebesta n.d, 13). Despite U.S. objections to its geographical footprint and services, in 1977, under the aegis of the E.U. Treaty of Rome, CEPT signed the Eutelsat interim agreements. Then in 1979 the Europeans established Inmarsat, offering a maritime communications service. The Europeans also competed directly with Intelsat in their attempts to create an African satellite subsidized by the European Union to serve former French and British colonies (Lloyd 1983).

The Europeans collaborated to produce state-owned, direct-to-home broadcasting satellites (Hills with Papathanassopoulos 1991, 102–4). Clay Whitehead, whose memorandum to the FCC had started the "open skies" policy, unsuccessfully attempted in 1984 to establish Coronet, a privately owned U.S. television satellite company in Luxembourg. That attempt then led to the Luxembourg-financed European Satellite Corporation (SES) and to SES rather than Coronet or intergovernmental Eutelsat becoming the major satellite television distributor (Marsh 1984; Snoddy 1985). Other regional and national systems such as Arabsat (1985), Brazil (1985), and Mexico (1985) took these markets away from Intelsat.

Intelsat also lost its technological cutting edge. Developing-country members insisted on retaining cheap C-band technology. The more directional, higher waveband Ku-band transponders were not introduced until the 1980s, and it was 1990 before Intelsat VI included spot beams allowing customized service to large users. Intelsat's failure to look after large users until 1984, or to appreciate growth in the Asia Pacific, gave a market to other operators with Ku-band satellites (Goldstein 1992, 12).

By 1985, despite receiving $3.4 billion from Intelsat contracts over the previous twenty years, the U.S. satellite industry claimed diminishing returns. The FCC calculated that of total annual sales of $6 billion in 1985, the industry would gain only $100 million from Intelsat contracts (*Telecommunications Reports* July 29, 1985). The figure appears to have been an underestimate. In fact, the first completely non-U.S. satellite (a Matra Marconi spare) was not commissioned until 1996—when Comsat was ordered to vote against the purchase (Weinberg 1997, 40). In addition, NASA launched Intelsat satellites until 1986. Only after the destruction of the space shuttle Challenger did the European Ariane dominate the launcher market (Whalen 2002, 161).

The End of Intelsat's Regulatory Function

The first indication that the Reagan administration was prepared to bypass Intelsat's regulation of international communications came in 1981. A letter to the FCC from Under-Secretary of State for National Security James L. Buckley said that there might be exceptional circumstances when it would be in the interest of the United States to use domestic satellites for international services.[66] Relying on this letter, the FCC decided that it had the authority to authorize "transborder" service—between the United States, Canada, Central America, and the Caribbean. The State Department and the FCC felt that approval of these applications by Intelsat's Assembly of Parties was not required. Instead, if the U.S. government and the foreign government concerned agreed in good faith that there would be no economic harm to Intelsat, then only technical coordination with Intelsat was necessary (Naraine 1986, 275).

Following on from this decision to allow domestic satellites to serve "transborder" markets, the wider international market became a target for the U.S. satellite industry. In 1983, Orion Satellite Corporation applied to the FCC for a private satellite over the Atlantic. Five other applications followed.[67] The congressional debate made clear that as well as being an outlet for the U.S. satellite industry, private satellites were perceived as transmission mediums for U.S. television programming and financial services into Europe.[68]

President Reagan signed a presidential determination in the "national interest" in November 1984. Private operators would be allowed to lease or sell their transponders to business users, thereby competing with Intelsat for the large-user leased-line traffic—about 15 percent of Intelsat's business. They also had to have a partner in another country before starting service. These, then, were to be U.S. international "enhanced" service providers.

The U.S. justification for its action in permitting competition to Intelsat originally rested on the distinction between "public telecommunications services" over which the treaty gave Intelsat regulatory control and "specialized" services that had only to be coordinated technically with Intelsat. Such a definition would allow the provision of services via leased lines to be termed "specialized" services and therefore unregulated. The model was similar to that used in 1972 by the Pentagon when seeking joint use of Intelsat satellites and mirrored the "specialized arrangements" of the ITU. But public telecommunications services were closely defined in Article I(k) of the definitive agreements to include fixed or mobile telecommunication services and "leased circuits for any of these purposes." The justification for the new operators had therefore to rest on the old argument that they would do Intelsat no "serious economic harm" (see Hills 1986, 166–79).

Intelsat continued to fight (Dunne 1986). Seventy-eight Intelsat members wrote to the U.S. government in protest. Congress accused it of a dirty-tricks campaign. Critics allege that the threat of separate systems empowered developing countries within Intelsat, which responded with programs such as its 1983 thin-route communications Vista service, Project Share (Satellites for Health and Rural Education) in 1984, and a development fund in 1985 (Intelsat 1986). This increased attention also united opposition to the private satellite entrants. By 1988 only Peru had availed itself of PanAmsat's promises of cheap transmission.

In 1985, faced with spare capacity, Intelsat began to sell or lease unused transponders for domestic telephone service at cut-price rates on a non-preemptible basis (Dunne 1985). An enraged Rene Anselmo, the founder of PanAmsat, accused Intelsat of predatory pricing. In addition, Intelsat began to modify satellites under construction and revised its earth station standards so that smaller antennae could be used (Colino 1986). Also, in view of the region's economic growth, from 1988 it began to put increased satellite capacity over the Pacific Ocean,

But then Richard R. Colino, Intelsat's chief executive, and his deputy, José L. Alegrett, were convicted of defrauding Intelsat. In December 1986 the State Department ordered Comsat to nominate previous FCC chairman Dean Burch, a close friend of Vice President George Bush, to replace Colino. Overt opposition from Intelsat to the new entrants collapsed.

In effect, Intelsat's regulatory function ceased with the Reagan decision of 1984, and so FCC protection of its traffic diminished further. In 1985 the FCC exempted all companies but AT&T from its circuit distribution guidelines and in 1988, following the entry of MCI International Inc. and U.S. Sprint Communications Company into the international market, it lifted its distribution policy in favor of a worldwide agreement between AT&T and Comsat.[69] In 1992 the FCC passed an order sunsetting the restrictions on the interconnection of private satellites with the U.S. public network from January 1, 1997. And in 1995, following pressure from U.S. satellite operators, it proposed lifting the divide between domestic and international satellite services, a policy that would allow U.S. operators to fan out abroad (Foley 1996a, 15).

Meanwhile, as FCC policies opened up the transatlantic market to new entrants, the major transatlantic revenue source for Intelsat diminished. Optic fibers, introduced into international communications in Europe in 1985, were lighter, easier to lay, more secure, required fewer repeaters, and also expanded bandwidth, so that one fiber pair could carry 4,000 circuits with up to four pairs per cable. In 1985 the FCC once more acted unilaterally, this time to allow private optic fiber on international routes. Cable & Wireless immediately set out to lay a Global Digital Highway from the United Kingdom to southeast Asia—across the United States (Cable & Wireless 1989). However, AT&T and a consortium of European PTOs laid the first transatlantic optic fiber cable (TAT-8) in 1988, carrying 40,000 telephone conversations.

Intelsat then launched its VI series in 1989, each satellite carrying the equivalent of 120,000 two-way calls. It claimed that the cost of a circuit year on the satellite was $504 compared to $1,596 on TAT-8. However, still preferring cable, most European PTOs and AT&T priced 56kbps and 64kpbs data circuits on TAT-8 at less that those on Intelsat (Hansell and Putney 1989, 44).

With declining fiber costs, non-operator investors entered the market. Investment increased from $500 million in 1988 to more than $2 billion annually from 1992, and new technology doubled the capacity of existing fiber cables. In an increasingly competitive market, owners refused to lease to competitors, thereby increasing the fiber installed (Foley 1996b). By 1995 an estimated two thirds of capacity on both the transatlantic and transpacific routes was idle, a proportion expected to rise to 80 percent by 1997 (Finnie and Schenker 1995, 1 and 9). Similarly, satellite capacity was set to rocket so that the 2,628 transponders owned worldwide by others than Intelsat was expected to more than double by the turn of the century (Foley 1996c, 8).

In 1989 Intelsat served 173 countries, carried 116,000 voice channels and 69,000 hours of television, and had 35 members relying on it for domestic services. The annual cost of space segment utilization had declined, but whereas from 1970 to 1975 the decline was 54 percent and from 1975 to 1980 it was 40 percent, from 1980 to 1985 it fell only by 8 percent, and from 1985 to 1988 by only 6 percent (*Transnational Data and Communications Report* 1989, 5). By 1991 Intelsat's revenues from public switched services had declined to 60 percent of the total (Intelsat 1991). Its potential decline then threatened, with unlimited liability, PTOs with large stakes (such as BT); pressure for privatization mounted. Eventually, unable with its existing organization to achieve the flexibility of competitive response that it required against intermodal and intramodal competition, Intelsat privatized itself in 2001 (Hills 2007).

Conclusion

This chapter demonstrates once more how international telecommunications was perceived from the U.S. side during this period as an extension of domestic communications and therefore subject to unilateral regulation. Despite the genuflection to international peace and understanding, Comsat was created as a U.S. public utility corporation with personnel determined to have an end-to-end U.S.-controlled satellite system. It was only the refusal of CETS to negotiate other than as a group that pressed Comsat into allowing joint ownership of Intelsat and an internationalized board of governors. Just as the State Department wielded little influence previously over the FCC, it was weak in its relations with Comsat. Intelsat remained entirely under Comsat control as it changed hats from manager to U.S. representative and manipulated the board's agenda.

It was Comsat's behavior that welded divergent views in Europe into an overall aim—to get rid of Comsat as manager. But until 1973 Intelsat had no juridical entity, so that it was Comsat and the FCC that had the direct power to issue contracts to primarily U.S. manufacturers and the FCC that regulated it as a U.S. carrier. In turn, that domestic regulation reinforced the congressional view of Intelsat as a U.S. corporation. The Europeans had little countervailing power until U.S. behavior and French determination brought them together to create their own satellites and launcher.

In the period of the 1960s and 1970s, it was AT&T's need for capital projects to expand its rate base that drove U.S. international communications policy. Its opposition in the 1960s to President Kennedy's bill ensured that Comsat served the carriers and that the FCC, rather than the State Department, was regulator. By refusing to accept other than joint ownership of earth stations, the carriers prevented bypass of cables by satellite. Then in alliance with the FCC, and by utilizing the Europeans' loathing of Comsat, AT&T laid cables of ever-larger capacity with an ever-larger contribution from the U.S. customer. Licensing was virtually the only mechanism used by the FCC to lower AT&T's tariffs, but those tariffs were based on an inflated rate base. The FCC never challenged the historic costings of AT&T that formed the basis for its high international tariffs, and only once its overinvestment in submarine cables. Only the Nixon administration publicized the cost advantages of satellites and attempted to restrain AT&T from undermining Intelsat at the expense of U.S. customers.

Because Comsat, as a private corporation, was the U.S. signatory, U.S. domestic politics prevented Intelsat from becoming an intergovernmental agreement on a one nation–one vote basis. Nor would the United States accept a division between Intelsat's regulatory and operational powers that would have undermined its monopoly. Although the United States was forced to accept an assembly, the assembly's weak powers meant developing countries had little representation. Because Intelsat averaged tariffs across thick and thin routes, FCC nonregulation of proportionate fill ensured that AT&T and foreign PTOs benefited at the expense of Intelsat's poorer clientele.

After the bilateral confrontations of 1974 with the Europeans, the FCC pulled back from collaboration and, excluded from North Atlantic planning meetings, Comsat and Intelsat went their own way. Overcapacity resulted. The 1980s U.S. response to large-user pressure was then to reintegrate national and international networks and unilaterally liberalize both cables and satellites. However, FCC attempts to export its domestic decisions on leased lines into the international cable market came up against the opposition of the ITU. Hence, when optic fiber cables came on stream, the FCC bypassed both the ITU and multilateral planning of transatlantic capacity to unilaterally allow private fiber cables.

Fiber optic cables then swung cost-effectiveness for long-distance point-to-point communications away from satellites, and overcapacity undermined Intelsat. Meanwhile, Intelsat's location in the United States made it an easier target than the ITU and made possible the Reagan administration's 1981 and 1984 unilateral decisions to directly override Intelsat's regulatory power. In turn, the 1984 decision extended the Nixon policy on domestic satellites that allowed the nonregulation of data transmission. The FCC's 1980 Computer II decision on the division of "enhanced" and "basic" services was transported into the international satellite sector. Eventually, end-to-end, company-to-company transmission could take place both via private satellites and via Intelsat. In effect, by virtue of U.S. operation of the eastern end of the most used and profitable international route, the FCC became an unacknowledged global international regulator. By the mid-1980s the FCC had begun to turn the clock back to end-to-end international networks under U.S. control. The next step was to liberalize ITU rules governing international cable networks.

3

International Market Structure
and the ITU

For almost a century, from its inception in the 1860s, the ITU had existed as a regulator of first the international telegraph and then international telephone networks. To the world outside the telecommunications industry this regulatory role was hidden in a "technical" organization. In reality, the ITU combined technical, economic, and legal aspects of the governance of the international communications system.

The previous chapter documented the end of Intelsat as a regulatory force. This chapter focuses on the domestic and international pressures, actions, and negotiations that led to changes in the ITU regulation of international market structure. The watershed was the 1988 ITU World Administrative Telegraph and Telephone Conference (WATTC) where a new set of regulations were arranged to replace those of 1973. The conference provided the forum for France and Spain to attempt to exclude "special arrangements" from the regulations altogether and for the United States to attempt to exclude the right of states to run their own networks. Both attempts failed. But the legal primacy of E.U. competition law in European member states signaled the beginning of the end of the "European state"-based monopoly market.

State-to-State–Based Networks: The ITU as Regulator

The Administrative Telephone and Telegraph Conference of 1973 had endorsed a radical revision of the telegraph regulations, cutting them from 151 to thirteen pages and the telephone regulations from thirty-five to eight pages. Both subsequently contained only general principles. In a move under discussion since the 1950s, the transfer of detailed regulations to CCITT recommendations allowed CCITT study groups to update them in line with technological change.

The CCI's work took place in study groups each addressing a particular "question" put forward by a member. Every four years the CCITT Plenary Assembly approved the recommendations on each "question" and agreed on study plans for the next cycle. CCI recommendations ensured continuity of technique and service, and produced international connectivity of basic services. By creating internationally compatible equipment, they also cut manufacturing costs (Purton 1987).

Although the ITU convention recognized the sovereign right of each country to regulate its telecommunications, the ITU recommendations reached beyond the international network. According to Jean Renaud (1987, 183) in 1949, the prewar convention that "no problem should be tackled by the CCIs unless that problem had international repercussions" disappeared from the texts. In legal application, ITU recommendations came below the regulations. Article I of the Telephone and Telegraph Regulations stated that "in implementing the principles of the Regulations," administrations (i.e., PTOs and recognized private operating agencies) "should comply with Recommendations" and "any instructions forming part of those Recommendations," including "on any matters not covered by the Regulations." Because they were mentioned in the regulations, CCITT recommendations were generally regarded as mandatory (Noll 1985, 27). In contrast, the radio regulations did not refer to the CCIR recommendations, which therefore lacked similar legal status.

But CCITT recommendations were not just about standardization. The multinational users of the International Telecommunications User Group (INTUG) could say in 1983 that "Study Group III of the CCITT deals with general tariff principles and through its position within the ITU and the United Nations system it is in effect the only existing truly international forum for determining international telecommunications policy" (Cullen 1983, 262). The State Department as well as AT&T represented the United States on this committee from 1956 (CCITT 1964–68, 29). It was Study Group III whose recommendations ensured that the structure of the international telecommunications market was based on a state-to-state system supported by a standardized system of tariffs and shared payments—"accounting rates."

As we saw in chapter 1, "special arrangements" which allowed bilateral arrangements to escape the regulations, were undefined. The British organized its intra-Empire preferential rates for public services under that clause, and the United States also argued successfully in 1949 that "special arrangements" included its companies' end-to-end networks and "regional agreements" (ITU 1949, Article 1). Both the British and U.S. systems were public telegraph networks. Although the British Commonwealth system did not formally end until the 1980s, the U.S. companies' end-to-end services faded with the introduction of TAT-1 in 1956. "Special arrangements" then became applicable mainly to private networks over leased lines.

Leased lines were first mentioned in the CCIF of the 1940s, but lack of capacity in the European telephone system prevented the leasing of circuits to private companies (CCIF 1954, Recommendation 21). Then coaxial cable technology provided additional capacity so that companies began to demand international leased lines on an end-to-end basis. But private networks threatened monopoly PTOs. A recommendation of the CCIF's last plenary assembly in 1956 stated that the leasing of circuits should not "be open to abuse by the subscribers renting the circuits." The calls over such circuits "must be concerned exclusively with the personal affairs of the subscribers or those of their firms. The line must in no way be made available to third parties" (CCIF 1957, Recommendation 21). The PTOs would not allow private competition.

By the mid-1960s this recommendation had been formalized into "Recommendation D.1" on the lease of international circuits for private service (CCITT 1965, 1–14). Although not allowing calls or signals originating with or addressed to third parties, the recommendation now allowed a "multiple-user lease" for companies "carrying on identical activities" or "active in the same field," but at a 37.5 percent surcharge. It specifically rejected interconnection into the public network and made it obligatory on PTOs to refuse service to resellers. These were virtually the same terms on which PTOs approved leased lines in domestic networks.

But when data communications became more widespread in the 1960s, the D.1 Recommendation caused problems. IBM complained of the high cost of leased lines for data transmission, on which there was a one-third surcharge.[1] The International Air Transport Association (IATA) wanted to be able to use leased circuits for a variety of traffic and wanted interconnection with the public network.[2] Despite the fact that AT&T had not signed the ITU Telephone Regulations, in 1966 the U.S. State Department asked for the expansion of the D.1 Recommendations, from "international" (European) to intercontinental circuits. Presumably, AT&T and the international record carriers wanted the recommendation to protect them from competition. But the State Department also asked that the interconnection of a leased circuit to the public network be allowed in order to "permit use of the leased circuit from the residences of customer employees" to take account of time differences. It argued that rather than taking away traffic from PTOs such interconnection would increase usage.[3]

In 1968, in response to these requests, the "D.2 Recommendation" was created to cover "continental" circuits. The D.1 and D.2 Recommendations were changed to allow flexibility of use and, although restricted to installations within the terminal country's boundaries, allowed interconnection into the public network subject to national regulation. In the United States, therefore, the interconnection of international leased circuits into the public network was possible eight years before the FCC ruled similarly on domestic leased circuits.

It was during the 1970s that data communications began to expand faster than voice. The Society for Worldwide Interbank Financial Telecommunication (SWIFT), the worldwide banking private network, launched in 1977, and in 1980 the CCITT passed "Recommendation D.6," which provided an exception to Recommendation D.1. It allowed such specialized international networks to closed user groups whose needs were not being met by PTOs.[4] The airline global communications network SITA (Société Internationale de Télécommunications Aéronautiques) was also authorized under this provision.

In general, so as to reduce tariffs on local telephone service, AT&T and the European PTOs kept international tariffs high. Since the majority of users of international communications were large business and the transfer of revenue to local service benefited urban voters, the distribution of costs and benefits in the network held political advantage to governments. But the practice increased incentives for multinationals to demand leased lines charged on flat-rate (bulk) tariffs.

Then, the introduction of competition during the 1980s in the United States, Japan, and the United Kingdom altered power relations within national networks. Because 60 percent of traffic was provided by large companies, after liberalization these large users became the primary focus for dominant operators. In turn, large users gained power in the international arena.

CCITT recommendations were directly contrary to the interests of private companies wishing to extend their networks globally and to sell capacity to third parties. They also acted to support the defensive actions of European PTOs. By 1981 European administrations—particularly in Italy and West Germany—were restricting access to leased lines and were seeking to widen the restrictions through CEPT (Thompson 1981, 68–69). U.S. companies began to complain to the European Commission.

Because its transatlantic leased line tariffs were lower than those of other European PTOs, BT became a hub for the distribution to Europe of 60 percent of transatlantic traffic. Then a number of message-forwarding agencies took advantage of these price differences and began relaying traffic between third countries and the United Kingdom. As required under CCITT Recommendation D.1, BT wrote to the companies forbidding the activity. But the European Commission applied Article 86 of the Rome Treaty, which prohibited abuses by market-dominant undertakings and found that BT had infringed Article 90(2) of the E.U. Treaty by behaving in an anticompetitive manner. The British government did not appeal; rather, it welcomed the liberalizing conclusions. In contrast the Italian government unsuccessfully took the European Commission to the European Court (Mestmäcker 1985, 317).

The European Court judgment meant that the ITU international system—of monopoly administrations collaborating within the ITU to protect those monopolies—had no legal standing in communications within the European Community. But the European Union at that time comprised only eleven members, whereas the ITU comprised more than one hundred, so the impact of this ruling was not immediate.

Computer II and International Communications

Based on its 1980 Computer II decision, unregulated "enhanced" services were the basis on which the FCC sought to liberalize international networks. But "enhanced" service suppliers needed freedom to use leased lines, and ITU recommendations blocked international leased line usage. However, the recommendations applied to users, not to operating agencies, so designation of users as operators could sidestep the recommendations. Therefore, in 1986, the FCC issued a policy statement finding that U.S. enhanced service suppliers who offered "public correspondence" services were eligible for designation as Recognized Private Operating Agencies (RPOAs) (U.S. Dept. of Commerce 1990, 69). As RPOAs (i.e., operators), they would be entitled to interconnect internationally and could evade ITU regulations through the "special arrangements" provision.

Following pressure by IBM Corporation, the 1981 British Telecommunications Act copied the FCC's Computer II division of the market. However, in the United Kingdom the division was stricter: value-added services were defined as those involving "substantial elements" in addition to basic network services. These were first allowed only in the domestic network, thereby complying with the CCITT recommendations.

But the proposal for a joint IBM-BT Managed Data Network using IBM's proprietal SNA standard led directly to the British decision in 1985 to liberalize all data communications under a new Value-Added Data Services (VADS) license. "Resale" and shared use of leased circuits were not allowed, and VADS suppliers were exhorted to use Open Systems Interconnection (i.e., ITU) standards. But by the time of WATTC 88, under the influence of Cable & Wireless (then pursuing its Global Digital Highway), the British put the onus on companies to negotiate with partners in other countries. Its grounds were that the CCITT recommendations were capable of different interpretations. In effect, it liberalized international data services under the "special arrangements" provision of the ITU.

In the meantime, because the U.S. trade deficit with Japan was $13.4 billion in 1981 and growing, the Japanese were particularly vulnerable to U.S. demands. The Americans first began to pressure Japan to buy more U.S. telecommunications products in 1979, when it targeted the procurement policy of Nippon Telegraph and Telephone (NTT), the Japanese domestic telecommunications administration. The Office of the United States Trade Representative (USTR) first employed here the tactic later used in European countries—expanding the issue beyond Post and Telecommunications Ministries to those in Finance and Trade. Conflict between these institutions was then expected to produce domestic debate and movement toward liberalization.[5]

In Japan there had been a turf war between the Ministry of Trade and Industry (MITI) and the Ministry of Posts and Telecommunications (MPT) since the 1970s over the convergence of telecommunications with computing. In order to gain domestic status, the MPT used its gatekeeper position to the ITU and argued that the regulations and recommendations were rigid international rules. Hence, it imposed Recommendation D.1, through which Kokusai Denshin Denwa (KDD), the state-owned international operator, had a monopoly of international communications. In 1984 there were reports that the U.S. companies Control Data Corporation and Tymshare had experienced delays and restrictions imposed by KDD on leased lines (Grub 1984, 143).

Because of the turf war between the ministries, the distinction between "basic" and "enhanced" could not be used in the legislation liberalizing the Japanese market. Instead, the market was structured between owners of facilities (Type I operators) and users (Type II operators). Special Type II entities comprised those supplying networks that linked a number of large-scale and international third parties.

The Telecommunications Business Law of 1985 indicated that the international provision of voice and data networks by Special Type II entities would be allowed under Japanese law. But because the provision was antithetical to CCITT Recommendations D.1, D.5, and D.6, the ministry continued to apply the CCITT recommendations in KDD's favor. It delayed access to Value-Added Network Systems (VANS) operators by demanding the use of international rather than proprietary standards for data transmission (Newman 1993, 373).

Bowing to pressure from the United States in 1987, the Japanese amended the 1985 Telecommunications Business law so that those companies who leased lines from the Type I facilities providers were translated from the status of "users" of leased lines to "providers." Although the Diet stated categorically that no Special Type II entity must do anything in conflict with the ITU regulations, by becoming "providers" they then became equivalent to "recognized private operators" that could furnish service to third parties without conflicting with CCITT Recommendation D.1. Later this interpretation of the CCITT recommendations was recommended as a means to make "international VAN services possible under the current legal order of international telecommunications" (Hirobe 1987, 16).

Despite gaining de facto bypass of the CCITT D.1 Recommendations between Japan, the United Kingdom, and the United States, USTR was simultaneously arguing in GATT and the ITU that enhanced service providers should be regarded as "users," not "providers," and therefore not subject to regulation—the exact opposite position to that which U.S. companies had pushed for in Japan. The reason for the U.S. position stemmed from the Nixon policy on domestic satellites, from the Computer II decisions and the divestiture of AT&T in 1984.

Under the Nixon policy and under Computer II, enhanced services were unregulated. Then, under the terms of the AT&T divestiture, the capital costs of the public network were reallocated from inclusion in long-distance tariffs to access charges for entry to the local network—"cost-based" tariffs. Although the FCC originally intended that access charges should be levied solely on individual subscribers, subsequent congressional opposition resulted in charges being divided between those levied on subscribers and those levied on "providers" of service. If enhanced service companies were classified as "users" of leased lines, not "providers," they did not pay access charges.

In 1987 the FCC moved to alter this situation, claiming that enhanced service providers should be treated not as "users" but as "providers"—a change that would benefit AT&T and other carriers. However, the FCC action created dismay among trade officials then arguing international enhanced services should be unregulated. Under pressure from USTR, the FCC backtracked. In this case it was the proposed regulatory structure of the international market that froze U.S. domestic regulation (Hills 1989).

Leading Up to WATTC

The decision to revamp the telephone regulations was taken in 1982 at the ITU Plenipotentiary Conference in Nairobi. Here, Nordic delegations, supported by the Japanese and CCITT Director Irmer Burtz, put forward a resolution suggesting "a new regulatory framework to cater for the new situation in the field of telecommunications services." The Nordic delegations were concerned about the proliferation of private networks using proprietary standards and the problems of interoperability between public and private networks. The Japanese also wanted exclusive ITU jurisdiction—a sentiment possibly not unconnected to the ongoing pressure from the United States. The CCITT director wanted definitional tidying up (Rutkowski 1986b, 10). The only significant opposition to calling a World Administrative Telegraph and Telephone Conference to redraft the regulations came from the United States, where AT&T was then fighting to retain its monopoly.

One can see a PTO mentality behind the proposal—the new services created "problems" for the state monopolies so they needed to be brought into the state-to-state system. But there was evident reluctance on the part of the ITU to tackle the issue—the first slot for the conference could not be found until December 1988.

1984 Preparatory Committee

The 1984 CCITT Plenary Assembly established a preparatory committee to prepare a draft text. At this 1984 meeting the Swedish delegation attempted to bring the U.N. Universal Declaration of Human Rights into the ITU by including within the regulations an obligation to universal service on the part of national governments. West Germany made a similar proposal later in the WATTC Preparatory Committee (PCWATTC) proceedings but neither provision found its way into the official text. Later in the 1980s and 1990s, Pekke Tarjanne, then Director General of the ITU, and a number of nongovernmental organizations (NGOs) and academics were also to take up (unsuccessfully) the issue of a "right to communicate" as an alternative discourse to that of the "Washington Consensus" of neoliberal economics (Hamelink 1994).

The PCWATTC was open to any participant in the CCITT Study Groups and had to prepare the text for November 1987 (CCITT 1984, resolution

14). Its chairman was Francisco Molina Negro from Spain where the PTO, Telefónica (a private company) had a very centralized stance over telecommunications. One of the vice chairmen, Earl S. Barbely, was from the U.S. State Department; others came from the Central African Republic, Hungary, and Iran.

Difficulties arose because ITU regulations were about market structure, not technology. The PCWATTC was also an attempt to rewrite the Regulations and define their scope, rather than to amend the existing text (Butler 1987). There was disagreement over whether the regulations should continue to list the services to be covered. The United States contended that such a list would exclude unforeseen technology (*Telecommunications Reports*, May 4, 1987). It followed that if there was not to be a list, then the Regulations had to define the scope of ITU regulations in some other way.

In the first meeting, the U.S. and Brazilian attempts to limit the regulations to "public" networks fell in the face of general agreement that there was no technically meaningful distinction between "public" and "private" networks in an integrated digital network environment (Rutkowski 1986b, 13). At one point the U.S. delegation was successful in limiting the scope of the new regulations to services "offered to the public," but then unsuccessful when the term "public" was defined as including any body within the geographical territory of the member (*Telecommunications Reports*, May 4, 1987).

In general, most PTOs were anxious that the new regulations should cover Value-Added Network suppliers. France, in particular, argued that in a competitive environment, private networks should be subject to the same technical and operational requirements as public networks (Rutkowski 1986a, 71). But the Americans and the British were adamant that the regulations should not apply to "users."

The possibility of international regulation of enhanced service suppliers stood in contrast to the success of their lobbying within the United States. The 1986 Computer III decision, which made it necessary for the Regional Bell Operating Companies to make their networks transparent for interconnection by 1988 through "Open Network Architecture," tipped the regulatory framework in favor of the new entrants. However, an alteration in the boundaries of "basic" services illustrated how the division between "basic" and "enhanced" services was political, not technical.

Under the AT&T divestiture agreement, the Regional Bell Operating Companies had to go to the courts to alter their supply of services. In petitioning Judge Harold Greene's Court in 1986 to be allowed to provide "information services" (then defined as "enhanced" services), the Bell Companies had argued that they could best be provided by operating companies as part of "basic" service—the opposite of the U.S. position in WATTC. Subsequently, the FCC moved to redefine such services as "protocol conversion" and "message forwarding" as "basic" service.

At this time the British had just put in place their VADS license intended to encourage new market entrants. After consultations with the industry, Oftel had accepted the use of proprietary standards (*Oftel News,* March 1986, no. 2). Hence, the British position within PCWATTC favored the use of proprietary standards in enhanced services. The British were also not prepared to see CCITT recommendations elevated into regulations and they had no desire to see the ITU increase its role and status vis á vis national regulation.

Final PCWATTC Text

Despite the ITU's normal practice of achieving consensus, U.S. and British interests were so hostile to the PCWATTC text that the final March 1987 report included annexes with adversarial comments (CCITT 1987). The major points at issue were that the text extended the regulations to "any entity . . . using the international network" and excluded the "special arrangements" clause (*Telecommunications Reports,* August 10, 1987). To gain alterations both Earl Barbely and Thomas J. Ramsey, the deputy bureau director responsible for ITU affairs in the State Department, were in favor of bilateral negotiations with key countries at "higher levels of government" (cited in *Telecommunications Reports,* May 4, 1987).

However, there was by no means unanimity between the various U.S. economic groups and administrative agencies. In the first round of comments to the FCC on WATTC, both the Association of Data Processing Services Organizations (ADAPSO) and IBM found the draft WATTC text "unacceptable."[6] Electronic Data Systems Corp. and ADAPSO wanted the United States to refuse to negotiate from the text. IBM and several of the Bell Companies made similar, but less aggressive, comments (Drake 1988, 226). Their fear was that the new regulations would place them in a more disadvantageous position than the 1973 regulations, which included "special arrangements."

Only AT&T and Nynex Corporation supported the effort to devise new international regulations. While AT&T accepted that some changes in the draft were necessary, it was primarily concerned that attempting to make major changes might isolate the United States. Similarly Nynex wanted the United States to work toward consensus (*Telecommunications Reports,* March 21, 1988).

Although U.S. regulatory structures were moving in favor of large users, technology was moving against. They feared that the development of Integrated Services Digital Networks (ISDN), bringing together voice, data, and images in one pipeline, would allow PTOs to preclude leased lines (*Telecommunications,* December 20, 1986, 24). These companies, represented by INTUG, wanted WATTC to give them both the right to establish private networks and the right of access to leased lines—an approach mirrored within the contemporaneous GATT negotiations (see chap. 6). What they demanded was the right to end-to-end establishment of private networks.

In the reply round to FCC docket 88-55, in an unusually forthright statement, AT&T opposed the views of the new entrants, stating, "The plain fact is that member governments that are determined to regulate value-added and enhanced services provided within their borders have the authority to do so regardless of the ITU regulations (*Telecommunications Reports,* March 21 and April 4, 1988). AT&T's views were evidently correct and were later supported by INTUG. Even liberalizing governments, such as the United Kingdom, had introduced licensing procedures for VADS operators that imposed very substantial conditions. Nevertheless, other new competitors, entering the debate for the first time in the reply round (such as Panamsat), objected even to international regulation of public operators. The FCC docket procedure actually made the task of satisfying all the domestic interests more difficult.

By early 1988 the American position on WATTC had become fundamentally linked to trade liberalization. In 1987 for the first time a negotiator for telecommunications, Carol Barlassa, was appointed to the Office of the Trade Representative. Some of those involved in WATTC were the same people previously involved in bilateral negotiations to open the Japanese market. Hence, opinion hardened against PTOs protecting their monopolies through WATTC.

In general, the principal concern of the United States and its allies was that the new regulations expanded the ITU's regulations to domestic regulation. However, while the PCWATTC text was oriented to reinstating PTO's overarching authority over international and national communications, the American alternative, by demanding rights for private competitors, attempted to impose liberalized regimes on other nations. In this debate the European Union was a natural U.S. ally, but crucially the European Community was handicapped by internal dissension and lack of formal status within the ITU.

Created in 1957 with five countries, the European Union expanded in the 1970s with the addition of Denmark, Ireland, and the United Kingdom. Greece joined in 1981 and Spain and Portugal in 1986. The European Union existed as a marginal institution of the 1960s and 1970s but gained momentum toward integration due to recession in the 1980s and the prospect of U.S. and Japanese economic dominance. In the belief that fragmented national markets could not give the necessary size or homogeneity for economies of scale, in 1985 the E.U. governments committed their countries to the establishment of a single European market by December 1992. Within this internal market, digital telecommunications was intended to increase the productivity and competitiveness of European manufacturing and retailing. Telecommunications was also the only industry within the information-technology sector in which Europe had a positive balance of trade in the 1980s, albeit because procurement was controlled by monopoly PTOs and supplied by "national champion" manufacturers such as Siemens AG of West Germany.

The E.U. founding document—the 1957 Treaty of Rome—is itself based on liberal trade principles, and the commission attempted during the 1970s to begin liberalization within the telecommunications sector by loosening the cartels of operators and manufacturers that controlled telecommunications procurement. But it was faced with such hostility that the sector was not included in its 1976 directive on the liberalization of public procurement processes. The European Commission instigated the creation of INTUG in 1974 to represent large users "after an E.U. spokesman complained that the Commission was deprived of telecom user input as it could not treat with national user groups without treading on the corns of member states, and no international group existed" (McKendrick 2000).

European PTOs were thought to have "successfully sidestepped" the commission through CEPT, whose twenty-six-country membership was much wider than the European Union (Solomon 1984, 220). Nevertheless, during the early 1980s, large users found that they could effect some liberalization by making a complaint against CEPT under the Treaty of Rome. But there were major differences in national regulation leading to frustration, not only among large users, but also in the European Commission (Solomon 1984, 220).

Eventually by linking telecommunications to the more politically salient information technology equipment sector, in October 1984, after four years of discussion, the commission gained agreement on the harmonization of networks and services and the opening up of 10 percent of national telecommunications markets. But the provisions had little liberalizing impact until the agreement on the single European market in 1987 gave the Competition Directorate, DGIV, more power in its debates with the dirigiste, DGXIII, the Telecommunications Policy Directorate (Hills with Papathanassopoulos 1991, 119–44).

The E.U. Green Paper

Until 1987 and the publication of its Green Paper on telecommunications, the European Commission made no further attempts at liberalization (Commission of the European Communities [CEU] 1987). INTUG later stated, "It is not an exaggeration to say that the Green Paper of 1987, which committed member states to deregulation, was largely the work of INTUG and a few other user organizations. There was nowhere else the CEU could go for informed knowledge to support the cause of deregulation" (McKendrick 2000). The User Group demanded more choice in services, cost-based tariffs, flat-rate leased lines, mutual recognition of standards, and a customer-oriented infrastructure, but not necessarily privatization. It also wanted PTOs to keep out of value-added networks, particularly data networks (McKendrick 1987). Data networks were still seen as the province of the U.S.-based computer industry, particularly IBM.

Nevertheless, because of the problems of definition, the E.U. Green Paper did not adopt the U.S. division of "enhanced" and "basic" services, instead preferring a division between those services "reserved" to PTOs and those not. Although the Green Paper supported some liberalization of data services and open network protocols, it was still predominantly concerned with the promotion of a European telecommunications equipment market through European standardization. DGXIII, concerned with European manufacturing, was not in favor of supporting U.S. company protocols for private operators. So, whereas DGIV favored the U.S./U.K. position, DGXIII did not.

Herbert Ungerer of DGXIII was to say later, "Since the Member States of the Community [were] individually members of the ITU, it was important to ensure that the positions adopted at WATT-C should be compatible with the Community's own framework for telecommunications regulation, as set out in the Green Paper of 1987" (1989, 2). The climax of a process of gaining consensus was the June 1988 European Council resolution on the development of the Common Market for Telecommunication Services and Equipment, with which the ministers representing telecommunications for the twelve member states adopted the policy goals of the Green Paper (Council Resolution 88/C257/01).

Ungerer (1989, 4) admitted that, because the PCWATTC text had been "worked out at a time when most countries had not yet started telecommunications reform," the consultations were difficult. Robert Priddle (1989, 5) the head of the U.K. delegation was to comment that "even within the European Community . . . it proved impossible in the lead-up to WATTC to agree on a consensus text." For the United Kingdom "[t]he need for some form of international regulation to ensure interconnectivity was not in question, but it was essential that such regulation should not limit the sovereign right of each individual member of the ITU to develop the national regime it deemed appropriate" (Priddle 1989, 2).

This question of national sovereignty divided Britain's Conservative government from other European member states in the run-up to the Single European Act of 1987. Prime Minister Thatcher's reluctance to cede sovereign rights over policy to the European Union created conflict between those whose vision was of an integrated federal entity with power skewed to the center and those, like Britain, who saw Europe's future as a federation of sovereign states. The United Kingdom was not simply concerned that France and its allies would railroad the commission into a restrictive stance to WATTC, but that the commission would use WATTC to extend its negotiating power. Whereas the European Commission held formal negotiating power on behalf of all member states within the GATT negotiations, it only had observer status in the ITU. Neither the United Kingdom nor France was about to cede their votes at the WATTC Plenipotentiary to the European Union.

Whereas the United Kingdom saw the PCWATTC text with its application of regulation to "any entity" as a great extension of the scope of earlier ITU regulations, for the French there was a real problem in the British/U.S. idea that some networks should be regulated less than others. After renovation during the 1970s, the French state network was the first to use the techniques of time division multiplexing and packet switching, and by 1989 it had the world's largest data switching network and the leading videotext service, Minitel. The French argued that a packet switched data network was no more an "enhanced" service than voice and had no desire to allow new competitors to take it over. In 1989 the French telecommunications regulator, Bruno Laserre, was to say, "If the public service has to connect everybody everywhere, if it costs money and the competition can cream-skim all that is profitable, that is the end of the public service" (cited in Dawkins and Dixon 1989). There was therefore a disparity between the EU's Green Paper with its division between "reserved" and other services and the views expressed by France and Spain.

The Butler Text

After the PCWATTC text had been circulated to administrations for their comments and amendments, a flurry of activity ensued. According to INTUG, "informal discussions took place between a number of members of the Union to find a common position that could be widely accepted, and . . . the Secretary General of the ITU was asked to provide facilities for one of these meetings" (Allen and Nicholas 1989).

Acting outside the normal ITU process the secretary general of the ITU, Richard Butler, invited nineteen of the most active participants in the pre-WATTC activities and the most vocal critics of the draft regulations to an informal meeting in April 1988. This was followed by another, larger gathering after an administrative council meeting (*Telecommunications Reports*, March 14, 1988). Faced with the prospect of the ITU losing its major donor—the United States—Butler also approached the European Union to gain backing for a more liberal approach to the PCWATTC text. Out of these 1988 meetings came an alternative text.

This "Butler" text altered the proposed regulations in order to allow private networks between consenting nations, to restrict the regulations to those international services "offered to the general public" (American terminology exempting leased lines), and to withdraw the obligation on users of leased lines to comply with CCITT recommendations. In particular, the "Butler" text excluded the most controversial article (Article 1.7) and created a new article (Article 9) that once again allowed "special arrangements."[7]

U.S. diplomatic activity in favor of the "Butler" text particularly to those countries with important trade relations with the United States followed. But this alternative "Butler" text had no official standing. Administrations had to comment and make alternative proposals based on the PCWATTC text, utilizing the "Butler" wording in their own proposals. By the time of the WATTC conference it was thought that the Americans had the support of the Nordic countries, the United Kingdom, Japan, New Zealand, the Netherlands, the Soviet Union, China, Colombia, and Brazil. Also, despite the lack of unity within the European Union, the presence of the Competition Directorate as an observer, together with the presence of ten representatives of INTUG—many within industrialized country delegations—were expected to boost U.S. lobbying power.[8]

But, despite U.S. diplomatic efforts, neither the United States nor the United Kingdom, nor the E.U.'s DGXIII appeared to have lobbied the seventy African, Pacific, and Caribbean countries that were signatories to the Lomé Convention.[9] Rather, at the plenary, the United Kingdom was put on the defensive by thinly veiled attacks on Cable & Wireless's monopoly in the Caribbean and it became evident that French expectations that Francophone Africa would vote with them would be disappointed. The wooing of developing countries to the "Butler" text seems to have been left to the ITU secretariat, placing the Secretary General in a vulnerable position in relation to his developing-country majority electorate.

The United States sent the largest delegation to the conference—thirty-seven members, of whom thirteen came from private industry.[10] In May 1988 the head of the U.S. delegation was announced as Earl Barbely from the State Department. However, Arthur Latno, a previous Pacific Telesis executive, then became ambassador in the State Department. He took over leadership of the U.S. delegation. Arriving in Melbourne without even a full set of papers, the U.S. leadership proved incompetent at public speaking, provoking embarrassment among its own delegation and derision from others. Inability to match the fluency of other delegations headed by career civil servants was compounded by silence at crucial moments. For instance, several requests from the chair for a speaker to propose the U.S.-sponsored "Butler" clause to reintroduce "special arrangements" were met by continued silence, forcing the British to respond. Delegate selection and the fragmentation of leadership then had a part to play in U.S. failure.

WATTC-88: The Conference

Refusing the use of the PCWATTC text as a reference text, the ITU secretariat put the "Butler" text before the conference. Since the PCWATTC text had come out of the initial CCITT plenary resolution, it is not clear how far this refusal was legal. However, this ploy in favor of U.S. interests created outrage among developing country delegates and distrust among the proregulation industrialized countries.

The result was confusion. The plenary session split into three groups, one to deal with accounting matters, a second to align the old regulations, and a third—an ad hoc group—to deal with the contentious clauses of Article 1 in the PCWATTC text and Article 9 in the "Butler" document. In the accounting committee, the chairman from Japan took the PCWATTC text as the base text and took amendments on that text, whereas, in the ad hoc group, the West German chair (himself a compromise between U.S./U.K. and French interests) refused to allow the PCWATTC text to be used as the base text. The secretary-general's composition of a drafting committee was then subjected to challenge as it left out members who felt strongly. On a number of occasions, the secretary general intervened directly to address members.

Although there was a tendency on the part of the industrialized members to see the position of the less industrialized countries as rigidly in favor of the PCWATTC text, without legal opinion available to them at the conference, their delegates argued that they had to support the original text. For some their support of the PCWATTC text originated from a system of administrative law demanding greater specification than that of common law; for some their positions were determined simply by distrust toward the United States and the United Kingdom and for some their support originated from considerations of PTO survival (Raveendran 1989). In addition many of the position papers had reached developing-country delegates only in Melbourne. As a result, editorial changes presented by the industrialized countries tended to fail in favor of the PCWATTC text.

Article 1.7 and Special Arrangements

The two articles most concerning the United States and its allies were Article 1.7 of the PCWATTC text and Article 9 of the "Butler" text. Article 1.7 stated:

> Members shall endeavour to ensure that any entity established in their territory, using the international telecommunication network to provide international telecommunication services:
> a) is so authorised by the member
> b) complies with these regulations, and
> c) to the extent considered appropriate by the Member complies with the relevant CCITT Recommendations.

The article thereby extended national regulation to users of the international network and to private networks and would allow PTOs to impose CCITT Recommendations on VANs. On the opposite side, the "Butler" text replaced Article 1.7 of the PCWATTC text (above) with Article 9, which read:

> Members may authorise their administrations, recognised private operating agencies, and subject to terms and conditions applicable under national law, any other organisation or person, to enter into special mutual arrangements with administrations, recognized private operating agencies, or other organizations or persons so permitted by national law in another country, for the establishment of special networks, systems, or applications, including the underlying means of telecommunication transport, to meet their own international communication needs or those of others who may use such networks, systems or applications.
>
> In making special mutual arrangements between administrations, recognized private operating agencies and other authorised organizations or persons, the parties concerned should take into account the relevant provisions of the CCITT Recommendations.

In effect this article not only defined "special arrangements," but it replaced the state-to-state basis of the international network of Article 1.7 with a potential multiplicity of private networks. It also made the use of CCITT recommendations (i.e., international standards) permissive, so creating the potential for a fragmented, incompatible, international network, or one based on a de facto standard established by market power—such as IBM's System Network Architecture standard.

"Butler" text supporters argued that Article 9 was procompetitive and Article 1.7 anticompetitive. However, the U.S. position was that Article 9 required further strengthening and should begin with the words, "These Regulations recognize the right of Members . . ." and that the second clause, referring to CCITT recommendations, should be deleted entirely. Following the hardening of its position between July and November 1988, the U.S. position was distinctly proliberalization and also anti-ITU de jure international standards.[11] Redrafted to its specifications, Article 9 would give the United States massive bilateral leverage to open other markets.

At the plenary the general approach of the "Butler" text was supported by the United Kingdom, India, Brazil, the Nordic countries, Bulgaria, and Japan. In addition, Canada, China, and Columbia wanted Article 1.7 suppressed. Those supporting Article 1.7 or wanting it strengthened included France, Greece, and Spain; smaller industrialized countries such as Belgium; and developing countries such as Mali, Niger, and Zimbabwe. France also had its own agenda of opposition to the United Kingdom. The United Kingdom was similarly antipathetic to France—opposing its chairmanship of a subcommittee. The Competition Directorate of the European Union, aligned with the United Kingdom, held meetings with E.U. member states throughout the conference and intervened to remind the French delegation that its actions were not compatible with the Treaty of Rome.

The Anglophone African nations formed their own bloc but grouped around themselves other developing countries on specific issues, including the French colonies. But there was no coherent bloc of developing countries. In particular Latin American countries tended to support the procompetitive position. Others, such as the Philippines, after an initial strong defense of PCWATTC, supported the U.S. position in public, while the Asian Newly Industrializing Countries, then under threat of U.S. trade reprisals, said little.

The poorer countries voiced vociferous opposition to Article 9. They were concerned that, whereas user groups such as SWIFT were exceptions to the 1973 regulations, Article 9 created the possibility of incursion into PTO revenues by major users. Despite an attempt by the United Kingdom to explain that these "special arrangements" would only be arrived at between consenting members and that Article 9 represented a counterbalance to Article 1.7, the African bloc was not appeased. Led by Senegal, they argued that "special arrangements" should take into account the "economic harm" that could be done to third countries, for instance, if private networks via satellite were to bypass their PTOs or multinationals were to choose to hub their communications outside their territory. As it became clear that neither Senegal, leading for the African bloc, nor the United Kingdom (leading for the United States) would give way, the plenary session adjourned to prevent the African group from walking out of the conference.

The major problem for the African group at this time was that it had no coordinated proposal to put to the conference. Eventually, with time running out on the hire of the meeting hall, the Australian chairman brought a package to the plenary and invoked the "Spirit of Melbourne" in order to produce consensus. Under this package, Article 1 would be approved without the U.S. amendment, together with Article 9, without any mention of "economic harm" as the developing countries had wanted. However, to placate them, the secretary general was to put a resolution to the ITU Plenary Assembly in Nice in 1990, drawing its attention to the economic harm that the "special arrangements" might do to developing countries. He would ask members whether a provision regarding such harm should be written into the ITU convention.

By this time it had become clear that neither the developing countries nor the allies of the United States were prepared to wreck the conference by rejecting the package. It seems unlikely that this information found its way to the U.S. delegation: the following morning, when the U.S. put its own unchanged liberalizing amendments to the plenary, it voted them down without a voice being raised in their favor. The U.S. delegation therefore ended the conference in a totally isolated position. The package was passed without the ITU's normal unanimity. The United States voted against and the United Kingdom abstained.

The Outcome

The final outcome replaced the pro-PTO text of PCWATTC with a "neutral" text. It watered down Article 1.7 from the PCWATTC position and balanced it with a weakened Article 9. The proregulation clause now read:

a) These Regulations recognize the right of any Member, subject to national Law and should it decide to do so, to require that administrations and private operating agencies, which operate in its territory and provide an international telecommunication service, be authorized by that Member.

b) The Member concerned shall, as appropriate, encourage the application of relevant CCITT Recommendations by such service providers.

c) The Members where appropriate, shall cooperate in implementing the International Telecommunications Regulations. (For interpretation, also see Resolution No. PL/2)

The article had to be read in conjunction with Resolution PL/2, stating that "upon requests by a Member concerned about the limited effectiveness of its national law in relation to international telecommunications services provided to the public in its territory," the members concerned are to "consult on a reciprocal basis" with a view to maintaining and extending international cooperation. The weak resolution was intended to placate developing countries concerned about their lack of ability to regulate competitive networks.

Article 1.7 no longer covered "any entity," whether user, service supplier, or operator. And rather than placing an obligation on members to ensure authorization, it recognized their right to do so if they wished. It therefore allowed liberalizing states to eschew a licensing scheme. And, rather than demanding compliance with CCITT recommendations, it placed the obligation on members to "encourage the application" of relevant recommendations. U.S. demands to have Article 1.7 cover only services "generally available to the public" (i.e., not leased lines) and to weaken the resolution to cover only "the international telecommunications network" (thereby excluding leased lines or private users) were not included in the final package.

In the Article 9 regarding "special arrangements," two concessions were made to developing countries—that arrangements should be made only between consenting countries and that they should avoid technical harm to the operation of telecommunications facilities in third countries. The article now read:

9.1 Pursuant to Article 31 of the International Telecommunication Convention (Nairobi 1982), special arrangements may be entered into on telecommunications matters which do not concern Members in general. Subject to national laws, Members may allow administrations or other organizations or persons to enter into such special mutual arrangements with Members, administrations or other organizations or persons that are so allowed in another country for the establishment, operation, and use of special telecommunications networks, systems and services, in order to meet specialized international telecommunication needs within and/or between the territories of Members concerned, and including, as necessary, those financial, technical, or operating conditions to be observed.

Any such special arrangements should avoid technical harm to the operation of the telecommunication facilities of third countries.

9.2. Members should, where appropriate, encourage the parties to any special arrangements that are made pursuant to 9.1 to take into account relevant provisions of CCITT Recommendations.

Article 9 was to be read with the accompanying opinion (PL/4), which stated that special arrangements "should be made only where existing arrangements are unable to satisfactorily meet the relevant telecommunications need" and that "in allowing such special arrangements, Members should consider their effect on third countries, and in particular, to the extent possible within national law" should endeavor to ensure that any adverse effects on other members are minimized. The proposal of developing countries to restrict "special arrangements" to those situations in which telecommunications administrations were unable to meet telecommunications needs (as in CCITT Recommendation D.1) was thereby relegated to the less legally powerful accompanying opinion. Their concerns regarding technical and economic harm were also only mentioned in the opinion. Again, the United States wanted a further weakening of the opinion.

In contrast to other delegations that moved from a proregulation stance to one of neutrality, the United States kept to its original position throughout. Lack of compromise contrasted strongly with other delegations, such as Canada and Australia, who therefore became the "honest brokers" between the liberalizers and the developing countries. The U.S. leadership also created hostility by its filibusters of the conference, made on the grounds that it had mislaid the relevant papers, but assumed to be so that it could consult the delegation. Finally, even its friends could not support it.

How could this failure be explained? One possibility is that a number of private interests within the U.S. delegation wanted the conference to fail. As time became short there was a certain amount of glee among American private interests. There was also ignorance. When faced with the final vote, the U.S. delegation then argued that the United States would be governed by the 1973 regulations because the U.S. Senate would not ratify the 1988 regulations for some time. ITU legal counsel disabused the U.S. delegation of the legality of this strategy. Whereas the United Kingdom, Canada, and Australia emerged with reputations enhanced, the United States managed to end the conference totally isolated and humiliated and with its relations with developing countries damaged.

Despite the U.S. isolation, WATTC allowed private networks, provided the government at each end was in agreement. But the structure of the international market was still based on the state-to-state model, with the difference that end-to-end private and public networks that might include "resale" were now allowed. The prospect was therefore of liberalized international networks between Europe, the United States, and Japan, subject to national regulation, but of state-to-state networks elsewhere.

INTUG summed up the result of WATTC-88: "The new regulations confirm the primacy of national sovereignty and allow for a range of national regulatory options to be applied to the private operating agencies which are providing international services offered to the public. This reflects the existing position today, and does not constrain the future development of either more liberal, or more restrictive, regulatory regimes" (Allen and Nicholas 1989). In effect, by emphasizing the importance of national regulation, WATTC provided a foundation to the World Bank/ITU focus of the next eight years and the inclusion of the Regulatory Paper in the WTO Basic Telecommunications Agreement of 1997.

Post WATTC

Even though the regulations had been altered at WATTC, the CCITT Study Group III still had to change the D-series recommendations to make them compatible. The French contribution was a rerun of its controversial proposal for an "access charge," or surcharge, to be levied on international carriers accessing national networks. In the Revenue Sharing Committee at WATTC, France had argued that the charge was intended to allow the development of value-added services while, at the same time, helping to cover their cost to national networks. At WATTC the proposal had provoked direct confrontation with the United Kingdom, supported by the United States, and a written note to European delegates from DGIV of the commission pointing out that, if the charge were discriminatory, it would contravene the Treaty of Rome. The U.S. comment that "access charges to enhanced service providers were not something we would want to impose in the USA," illustrated how crucial it had been to get the reversal of FCC policy that would have imposed them.

Again at the CCITT Study Group III discussions in 1989 and 1990, the United States and INTUG looked for wording of Recommendation D.1 that would allow private companies to lease international circuits and do whatever they liked with them. France and its supporters acknowledged that the D-series restrictions were out of date but argued that safeguards were needed to protect the financial viability of public networks.[12] France gained support from Japan, where U.S. companies had to pay a surcharge of up to 20 percent for leased Japanese international facilities (Cowhey and Aronson 1991, 304n19). Japan wanted to keep that part of the D.1 Recommendation that forbade interconnection of private leased circuit networks at both terminals unless there was agreement between the two administrations involved. Canada wanted the carriage of traffic from third countries forbidden. Together, France and Japan wanted the use of CCITT standards within leased circuit networks.[13] Study Group III provided a rerun of the same arguments as WATTC-88.

However attitudes were changing: 1990 marked the beginning of U.S./Japan/U.K. International Value Added Networks that allowed shared use and resale of leased circuits for other than basic voice services (Cowhey and Aronson 1991, 304). The triad had also licensed multiple international carriers for voice services. In addition, AT&T and other carriers were setting up overseas subsidiaries or buying into them in order to collect business user traffic.

At the CCITT, the E.U. Competition Directorate (DGIV) became involved. It warned that restrictions on private networks might breach the Treaty of Rome's antitrust provisions.[14] Finally, at the end of 1990, Study Group III agreed to a revised D.1 Recommendation under which international leased lines could be shared between companies and could be interconnected into public networks at one or both ends. Also, in principle, these leased lines would be cost-oriented and flat-rate based, rather than charged according to the volume of data passed. To that extent, the United States and INTUG had gotten their way. But other administrations had ensured that the rights of states overrode those of multinational users. So the leased circuits could be interconnected into the public network only with the permission of both administrations. PTOs were also allowed to charge for any additional costs that they might incur in the provision of the circuits (as in the recommendations of 1957) and to exclude companies from the provision of specific services, such as the public telephone.[15] Hence the model accommodated those countries where liberalization had taken place but did not prevent monopoly networks where governments preferred—the state was still the basis of regulation.

So, for instance, the Japan-U.S. VANS agreement of 1990 allowed only intracorporate communications between offices of a single company or joint use where an ownership or business relationship existed. "Resale" was not permitted. The national government retained control (Meyers 1992, 11). The U.K. government demonstrated similar control when in 1992 the Director General of Oftel, the U.K. regulatory agency, issued licences for International Simple Voice Resale.

Both these and the 1997 Public Telecommunications Operator licenses that followed the full liberalization of the U.K. market contained a provision that applied to companies that were operating systems outside the United Kingdom. Under that provision, Oftel could take action where it appeared that a licensee had acted in such a way that "competition in the provision of any telecommunications service . . . in the U.K. is being or is likely to be restricted, distorted or prevented" (Oftel 2000, 24). In other words, the license conditions were extraterritorial in their application, and the regulatory agency could prevent any unfair competition by foreign companies at the overseas end of networks (the "Western Union" model could not reoccur). By 2000 there were fifty International Simple Resale Voice operators and twenty to thirty telecommunications operators active in the U.K. international market, but none could evade national regulation (Oftel 2000, 1).

Conclusion

This chapter began at a time when the international network was strictly under the control of national PTOs and recognized private operating companies. Through ITU regulations and CCITT recommendations, these national monopolies could exploit the international network to charge high tariffs to large businesses. But the increasing power of the European Commission, its alliance with multinationals, and its commitment to a single market began from the 1980s to erode monopolies within Europe. Domestic liberalization in the United States, Japan, and the United Kingdom also swung economic power to large users and digitalization brought with it new entrants, such as IBM, from the data processing sector. These new entrants came from a sector that, unlike telecommunications, had never been regulated.

Having originally tried and failed in 1980 to open up the international market to "enhanced" services, the United States was successful in leveraging the issue into its trade dispute with Japan. By determining in 1986 that VAN suppliers could be designated as "recognized private operating agencies," the FCC created the idea of how to bypass the ITU regulations. Subsequent pressure on Japan effected a similar change in that country's regulation.

The chapter demonstrates how U.S. domestic policy, promulgated by international large user groups and U.S. data processing companies' using their economic base in European countries, was instrumental in defining the structure of national and international regulation. The U.K. regulatory structure, where IBM held political influence, came to virtually mirror that of the United States. However, the end result of the WATTC negotiations did not meet the more extreme demands of companies such as Panamsat— that all international telecommunications services be unregulated. Nor, as some U.S. businesses proposed, did it create an "Empire rules" model that would prevent the national regulation of leased lines. It did not place multinational users over and above nation-states and in that it was a failure for some U.S. companies.

The case study of these multilateral negotiations shows how industrialized countries with sufficient personnel to staff meetings and sufficient legal expertise to understand texts held indirect institutional power over the proceedings. Techniques of late submission of papers, of informal meetings with the personnel of the institution, agenda manipulation by that personnel, as well as direct relational pressure on participants all play their part in achieving goals. With the organization under financial threat from the United States, the ITU director general acted in favor of the liberalizing industrial countries rather than the majority of ITU members. The failure of developing countries in these negotiations was due both to economic vulnerability and to lack of preparation of a "bloc" position. That they came away (in these circumstances) with little loss and some gain was due to the chairman, who gambled that the proliberalizing industrial countries would have gained enough in the final package not to wreck the ITU.

What the U.S. companies and Cable & Wireless wanted was the right to establish private networks, to sell leased capacity to third parties, and to operate end-to-end networks using proprietary standards of their own choosing. Such a scenario threatened fragmentation of the existing state-to-state–based international network by preventing global interconnection and by overriding national regulation. What emerged was the principle of state sovereignty intact. Although the previous rigid monopoly system was eased, national governments could subsequently point to the 1988 regulations to demonstrate their right to license foreign operators. That licensing could, as in the case of the United Kingdom, be extraterritorial so as to control the overseas behavior of the operators of end-to-end private and public networks.

Although the eventual revision of the CCITT recommendations allowed "special arrangements" that no longer excluded "resale," the hidden tradeoff was that that liberalization took place under strengthened state supervision.

The eventual package of rules both recognized and strengthened the principle of sovereignty. By doing so, it also de facto accepted the changes in the international market created by U.S. domestic regulation and its division between "basic" and "enhanced." But, because WATTC did not substantially alter the state-to-state–based system, U.S. companies transferred their attentions to GATT, as we shall see in chapter 6. Also, the final WATTC regulations, while stripping the ITU of one aspect of regulatory power supporting the state-to-state model, did not alter its power over international standardization or "accounting rates" or reflect the shift in power to large users. That is the story of the next chapter.

4

Markets and Membership:
The Restructuring of the ITU

The changes to ITU regulations that liberalized the state-to-state network system still left the ITU with its role in governing "accounting rates" as the central organization in standardization and with control over the frequency spectrum and orbital slots. The previous chapter demonstrated the inability of a U.S. delegation to function well within international multilateral negotiations. This chapter reviews the way in which the United States and other industrialized countries reacted to and created change within the ITU. It documents first how they prevented developing countries from gaining power in the organization's U.N.-based one nation–one vote structure. It then follows their demands for restructuring to allow private companies to exert greater influence in the organization.

The chapter explains how the ITU became useful to the United States, the World Bank, and other liberalizers for the indirect power it gave them. From 1991 the ITU's concept of development shifted from state-led modernization to the neoliberal primacy of markets. Its large membership gave it the status to influence developing countries. Nevertheless, the ITU found itself bypassed—by the FCC on global mobile personal communications satellites (GMPCS) and on "accounting rates," and by the Clinton administration on Internet regulation.

Restructuring the ITU: Developing Countries Gain Power

When the ITU was restructured under the leadership of the United States in 1947, it was the interests of U.S. operators and manufacturers that dominated. It was in the permanent CCIs and the newly created International Frequency Registration Board (IFRB) that the work of the ITU was to be done. Engineers dominated in a secretive organization, which banned the press from meetings.

The membership of previous colonial countries from the 1950s introduced telecommunications networks that were not at a similar stage of development and were outside the temperate zone. The first recognition of this change came in 1952 when "with some reluctance" the ITU plenipotentiary passed a resolution asking the United Nations Development Program (UNDP) to include telecommunications in the areas for which assistance was available (Codding 1989, 13). The distribution of UNDP funds became the primary development activity of the ITU for the next thirty years.

After a new clause was added to the 1959 Convention to include fostering "the creation, development, and improvement of telecommunication equipment and networks in new or developing countries," the CCIs began to develop handbooks for developing-country members and to devise general plans for international telecommunications networks in Africa, Asia, and Latin America. But the industrialized countries defeated proposals for a development assistance fund, for regional offices, and for a new section of the ITU for technical assistance on a par with the CCIs (Codding 1990, 140; 1989, 20). The proposals for regional offices and training centers came up against the Soviet Union's concern that these would lead to penetration by Western capital (Kazuka 1989, 52 and 56–57). Developing countries were caught in the East/West divide.

The 1973 plenipotentiary ended colonial voting, but Charles Kazuka (1989, 58), who was present, noted that "the industrialised nations . . . were not enthusiastic" about development assistance. Of thirteen proposals on the issue, the plenipotentiary agreed only to an Emergency Fund for Technical Cooperation financed by voluntary contributions and donations of second-hand equipment. In particular, the United States argued against the fund, stating that it would not contribute.[1] As a result of the south's demands, the U.S. delegation talked of withdrawal from the ITU (cited in Codding 1989, 16).

But the developing countries had a newfound confidence derived from the Organization of Petroleum Exporting Countries (OPEC), evident in the adoption of the New International Economic Order declaration by the U.N. General Assembly in 1974. In the ITU, they began to demand equity. According to William H. Read, who had served as an advisor in the White House Office of Telecommunications Policy, industrialized delegations regarded the demand for maritime frequencies by even landlocked countries at the Maritime World Administrative Radio Conference (WARC) in 1974 as "mindboggling" (cited in Kinn 1985, 44). Then, in 1976, Colombia formulated the "Bogota Declaration," claiming that equatorial nations' sovereignty extended to include the geostationary orbit—22,300 miles above Earth. The following year, at a broadcasting WARC, the United States was isolated in refusing to change from the traditional "first come, first served" approach to orbital/frequency slot assignment to an "a priori" allotment plan promoted by the South (Codding 1984, 439).

The United States was insensitive to these warning signals in preparing for the major 1979 WARC. Lady Diana Dougan was later to comment on the U.S. delegation that "although they were perhaps the world's best experts in the technology of telecommunications, they were not prepared for the Third World's shift to political goals" (cited in Kinn 1985, 46). There was a lack of respect. One Consumer Union delegate reported that at the first official meeting of the U.S. delegation, a member referred to "banana boat countries."[2]

WARC 79 was the first technical meeting where the developing countries were in a majority. U.S. demands for additional shortwave (HF) allocation for Voice of America brought it into direct conflict with delegations from the south, many of whom used shortwave for point-to-point communications. The difficult issues—high frequency, satellite orbits, broadcasting satellites—were postponed to further specialist conferences (Home Office 1980). But the United States alienated many. Rather than entering reservations to protect its existing high-frequency broadcasting interests, despite having lost the vote on the expansion of broadcasting, it affirmed its right to use those additional bands it wanted until the specialized conference took place; in effect, it ignored the conference vote (Valentine 1981, 141). However, according to George Codding (1984), the one item that "seemed to shock the United States most was the successful attempt . . . to elect an individual from the Third World as chairman, thereby overturning the previous 75–year process of informal agreement between the major industrialized countries" (Codding 1984, 444).

The 1982 Nairobi plenipotentiary marked a turning point as the first to be held in a developing country. In the previous ten years, seventeen new countries had increased the developing-country majority. Amid U.S. threats of withdrawal, the conference was nearly wrecked by attempts to expel Israel for its invasion of Lebanon. Eventually, in response to the seventy-six items placed on the agenda by the developing world, the conference altered the ITU Convention to include "technical assistance to developing countries" among the organization's goals. But again the proposal for a department within the

ITU devoted to technical assistance, equivalent in status to the CCIs, failed (Codding 1990, 140).

Some developing-country members suggested cutting back on budgetary allocations to the CCIs in order to make available more funds for technical assistance (Näslund 1983, 110). But because each country chose its level of financial contribution to the ITU, they had little leverage. The United States threatened that should the plenipotentiary agree to a development fund, it would simply decrease its "class-unit contribution and thus decrease ITU resources" (Doran 1989, 35). Instead Resolution 19 established a Special Voluntary Program for Technical Cooperation to bring in aid from industrialized countries based on voluntary donations from private industry. And, in order to head off further demands from the south, the United Kingdom, the United States, and Japan backed the establishment of a committee (the Maitland Commission) to study developing-country problems.

Within the ITU the technical allocation of orbital slots to Direct Broadcasting Satellites in 1977 had been based on the presumption that they would be domestic broadcasters. All but three states, one of which was the United States, opposed DBS international systems and approved the principle that broadcasting across borders required prior agreement from the host government. Then, when the forum changed to UNESCO, the United States advanced the same "freedom of information" arguments as in the 1940s for the same content industries, against Soviet-led support for state regulation (Naraine 1985). Defeated, the United States withdrew from UNESCO at the end of 1984 and Britain's Thatcher government followed suit in 1985.

There was a similar U.S. debate about withdrawal from the ITU. Following the Nairobi plenipotentiary, the United States cut its funding and ITU laid off staff. The U.S. Office of Technology Assessment "discussed options for changing the structure of the ITU, including seeking revision of the voting formula and increasing regionalization of the ITU so that only those nations most directly affected would participate in particular decisions." Other suggestions included "creation of some new entity, perhaps in the mold of Intelsat, for managing spectrum and orbit matters and achieving consensus on technical telecommunications matters" (cited in U.S. Congress, Senate 1983, 29).

But U.S. academics and others staunchly defended the ITU. Anthony Rutkowski found that 85 percent of all the major meeting days in 1983 were devoted to CCITT and CCIR work, of major concern to U.S. companies (1984, 29). And, as a positive response to the 1982 plenipotentiary, due to the leadership of Ambassador Michael Gardner, the U.S. Telecommunications Training Institute was established.[3]

This was the era of elite commissions such as the Brandt Report (1980) on the south. The ITU's Independent Commission of Worldwide Telecommunication Development (Maitland Commission) was composed of a "range of private sector experts—making it more acceptable to reluctant industrial countries" (Doran 1989, 36). Between the 1979 WARC and the publication of the Maitland Report (so named after its chairman, Sir Donald Maitland), a number of studies from the World Bank and from academic Heather Hudson highlighted the worldwide inequity in telecommunications provision (Saunders, Warford, and Wellenius 1983; Hudson et al. 1979; Hudson 1983; Hudson 1984). However, the Maitland Report's comment that there were more telephones in Tokyo than the whole of Africa became world famous. It drew attention to the disparity of telephone density between industrialized and developing countries and also to that between urban and rural areas within the developing countries (ITU 1985, chap. 2).

Despite the multiplicity of factors accounting for these disparities documented in the Maitland Report, the one that gained most favor among industrialized countries was the lack of priority afforded to telecommunications by developing-country governments. Along similar lines, the Maitland Report recommended the establishment of a Centre for Telecommunications Development within the ITU to act as a development policy unit, bringing together and disseminating information on the role of telecommunications in economic and social development (ITU 1985, 53). But the report did not specify where the funds for the Centre were to come from.

Subsequently, when the Centre for Telecommunications Development came into being in 1985 there was no ITU budgetary allocation for it. Nor was it very successful in attracting private money (Kerver 1989, 29–30). Donations were often linked to specific projects, such as the unsuccessful Regional Satellite for Africa (RASCOM) and, given the small size of the budget, tended to benefit Western consultants.[4] Companies' financial reluctance increased when donations funded the procurement of equipment and services from competitors (Bellchambers 1994, 34). Lack of resources, increased competition for funds, and "an insufficiently dynamic response from international organisations, governments and telecommunications operators" dampened the enthusiasm raised by the Maitland Report.[5] Although the developing countries had held a majority in the one nation–one vote plenipotentiaries of the ITU for thirty years, throughout this period the industrialized countries, led mainly by the United States, had managed to sidestep that voting power through financial clout and the tradition of consensus.

The Hansen Report: Liberalization, Privatization, and Regulation

Aware that the ITU required restructuring, Richard Butler, the ITU secretary general, in May 1988 appointed an advisory group on telecommunications policy to kick-start the process. The group consisted of individuals in a voluntary capacity from the World Bank, from the FCC, and from business and academe under the chairmanship of Paul Hansen of Denmark. Its remit was to undertake an overview of structural changes in the global telecommunications sector and to report on how governments and the ITU should respond.[6]

The ensuing 1989 report was wide ranging. The group argued that PTOs could no longer "provide the increasing diversity of communication services necessary to meet the expanding variety of communication needs and demands" (ITU 1989, II 2.1a). Cheapening technology was already allowing large users to bypass public networks. Its answer was increased commercialization of PTOs, "cost-based" tariffs, and liberalization of services—more or less taking the World Bank line at the time. It argued that developing countries needed national strategies, not only related to installing basic infrastructure, but comprising "such issues as sector restructuring, policy development, appropriate regulatory framework, private and foreign participation" (ITU 1989, para. 7.5).

In the first example of the influence of World Bank telecommunications personnel, the report suggested a "clear separation of telecommunication operations from government administration" (para. 4.3) and the introduction of regulatory mechanisms at national level. This was the first time outside the European Union that national regulation was given primacy in the transition from monopoly to competition. The report concluded that the existing structure of the ITU gave little attention to the harmonization and coordination of national policies and the development of networks and services in developing countries (para. 6.2). It indicated how the World Bank and ITU might operate in a new internationalized economic environment: "Major international agencies will have a growing role to play in at least two areas: a) helping to identify and assess a range of strategic options for developing countries, and b) establishing a policy framework for building constructive links between developing countries and the industrialized world" (para. 8.7).

These two roles were quite contrary to previous ITU goals. Where economic matters had arisen that were not in the industrialized countries' interests, the cry had always been that the ITU was a "technical" organization.

The ITU had also steered away from any direct interference with domestic systems. However, the report pointed out that expansion of telecommunications networks and services had pushed many issues of national policy to the international level. Global information and communication networks needed compatibility, not only of technical standards, but of national telecommunications policies, regulations, service offerings and tariff structures (para. 2.1b).

In the long term, the Hansen Report had a wider impact than on the structure of the ITU. A subsequent ITU report commented, "Probably the most important contribution of the Hansen Report was that, thanks to its objective articulation of the impact of the changing environment on both developing and developed countries, it no longer was 'heretical' for developing countries to consider introducing elements of liberalisation, competition, privatisation and foreign participation, subject of course to the sovereign right of the country concerned to choose the specific telecom policies, structures and regulatory system tailored to its particular requirements."[7] The Report marked an era of World Bank influence over the ITU. However, in the short term the report formed part of a campaign during 1988–89 by the outgoing secretary general, Richard Butler, to alter the structure of the ITU (Butler 1989a; Codding 1990, 143; Codding 1991, 283).

Butler wanted to reform the CCIs' standard-setting process so as to hold onto its industrialized membership. Amalgamation of the CCIs would not only save money but would also combat the new regional standards-making bodies—the Telecommunications Technology Committee in Japan, the U.S. T1 Committee, and the European Telecommunication Standards Institute. For the United States these regional standardization organizations provided a potential route to bypass the ITU—a proposal defeated by Japan. For Butler these agencies had the potential not only to supplant the ITU but also to fragment the world market into protectionist, regionally based groupings (Butler 1989b, 5–6; Besen and Farrell 1991, 311–21).

Following Butler's views, Ethiopia put forward proposals to the 1989 plenipotentiary for a new Telecommunications Development Bureau, the restructuring of the IFRB, and the merger of the CCIs (Codding 1991, 284). But the United States was against the merging of the CCIs and "implied that its future interest in the ITU would be reduced if it were enacted" (Drake 1988, 30). Despite the fact that the CCIR had had an average attendance of two people at its meetings over the previous ten years, the United States preferred it to the CCITT because it was less subject to political maneuvering (Finnie 1988). In addition, the United States wanted to retain the IFRB, which had traditionally had a U.S. director. The U.S. delegation's report commented that the proposed changes "were not developed in detail . . . and all represented significant change for which a rational presentation of the pros and cons was strikingly absent" (U.S. Department of State 1989, 14–15). Concerned about losing power within the institution, the United States supported the status quo.

The developing countries pressed the Maitland Committee's recommendation for a Telecommunication Development Bureau. They had discovered that while the six richest countries contributed only 30 units each—in total less than 50 percent of the budget—CCI work took up 68 percent of the budget. They concluded, therefore, that they were paying toward standards activities.[8]

A compromise between the proponents of change and its opponents led to the immediate creation of a new semiautonomous Telecommunication Development Bureau in return for agreement that the structure of the permanent organs (CCITT, CCIR, and IFRB) should remain unchanged in the short term.[9] Developing countries gained a development bureau with an equal status to the CCIs, an elected director, and $10 million from the ITU budget (Renaud 1987; Codding 1989). But the plenipotentiary also established a so-called High-Level Committee with representatives drawn from twenty-one mainly industrialized countries "to carry out an in-depth review of the structure and functioning of the Union."[10]

In a mark of changing attitudes, the 1989 plenipotentiary passed a resolution welcoming users' participation in policy and agreed to a previously unsuccessful U.S. proposal that scientific and industrial organizations could join standardization committees if "their" government agreed (Pipe 1989, 9). Previously, manufacturers of telecommunications equipment and user groups could only attend study groups on an advisory basis and were excluded from plenary sessions where the study questions and recommendations were made final (Codding 1993, 24). Now they could attend, under the watchful eye of governments—but not vote at—CCI plenary sessions.

The liberalizing countries, led by the United States and the United Kingdom, also succeeded in the election of Pekke Tarjanne of Finland as secretary general. Tarjanne's opponent, Francisco Molina Negro of Spain, the chair of the Preliminary WATTC Committee, was an enthusiastic supporter of state-to-state networks. In contrast, Tarjanne came to the job with experience of Finland's competitive system.

ITU Restructuring: The Rise of Large Users, Private Industry, and the World Bank

Pekke Tarjanne began his tenure with a belief in the principles of universality and the right to communicate as a fundamental human right (Tarjanne 1992a, 45). Yet, soon adopting mainstream Washington views on the irrelevance of nation-states, Tarjanne became the advocate of inevitable "globalization" and the primacy of capital over governments. He argued,

- Privatisation has reduced the competence of ITU member countries in dealing with technical issues, particularly in the area of standardization.
- Competition has reduced the importance of government's traditional regulatory role in controlling the industry and transferred power to the marketplace.
- Deregulation has become the new model for telecommunication development, displacing traditional approaches based on bi- and multilateral financial and technical assistance.
- Regionalisation has created new standardisation, regulatory and development structures between the nation state and the ITU.
- Globalisation may change the basic structure of the telecommunications industry so that it is dominated by a small number of global actors or alliances which extend beyond national boundaries (Tarjanne 1992b, 13).

Some commentators argue that from 1991, under Tarjanne's leadership, the ITU adopted the credo of "deregulation" and proceeded to "devote remarkable energy to spreading it" not just to the industrialized and the transitional economies of Central Europe, but also to the least-developed countries. They argue that there was no discussion of this change and no preliminary evaluation of the socioeconomic consequences, or of the impact on public service resources (Egger and Fullsack 2003). From the organization's perspective, if states were no longer important, power transferred to the private sector.

Even before the High-Level Committee report in December 1992, Tarjanne had appointed a World Telecommunications Advisory Committee made up of eighteen senior representatives of the global telecommunications industry. And, in response to technological convergence, he had warned that the ITU must enlarge its membership to include more private industry from computing and broadcasting and would have to eliminate the distinctions between member and nonmember participants. The implication was that companies should have votes equivalent to governments. A major reason was money: because government support was declining, the ITU needed to attract new sources of financing (Tarjanne 1992b, 13).

The High-Level Committee report, prepared with consultancy support, proposed that the work of the ITU should be reorganized into three sectors: Development, Standardization, and Radiocommunications. Each was to be served by an elected director, have its own budget, and hold its own World Conference supported by study groups. The old IFRB was to be downgraded and (with the CCIR) to become part of the Radiocommunications Sector (ITU-R). The Telecommunications Development Bureau would become the Telecommunication Development Sector (ITU-D). The CCITT and the standards-making part of the CCIR would become the Standardization Sector (ITU-T). The secretary general's overall coordinating role would be enhanced, supported by a new Strategic Policy and Planning Unit and a Business Advisory Forum "through which he could conduct a dialogue with business leaders" (ITU 1991, 269–70). In acceptance of the changed telecommunications environment, the additional plenipotentiary, called to accept the report's proposals, also created advisory boards from industry for each of the three sectors.

The overall effect of the restructuring, under which each of the three sections would have their own plenipotentiary, was to isolate developing countries within their own section of the organization, talking to themselves. In effect, the new arrangements represented the reality—that those who held the purse strings (the World Bank, commercial banks, large multinational users and manufacturers)—were now important to the ITU. And in recognition of these new realities, for the first time the views of the World Bank were put before the Additional Administrative Conference by the secretary general.

The World Bank's paper outlined its then financing of telecommunications based on structural adjustment, privatization, and liberalization. It also proposed a range of activities that the ITU should undertake in the development sector, including a panel of policy and regulatory experts, the preparation of medium-term investment programs giving quality assurance to the funding agencies, and the bringing together of funding agencies both multilaterally and within countries.[11] In other words, the World Bank foresaw cooperation in terms of the ITU's helping developing countries restructure their telecommunications sectors for foreign investment. In effect, the ITU would become an agent of the telecommunications sector of the World Bank. In particular, what the ITU could do that the World Bank could not was provide ongoing advice on regulation.

Within one year the ITU had initiated an annual "regulatory colloquium" funded by infoDev of the World Bank. The intention was to bring together industrialized and developing country personnel to discuss regulation at the national level. But of the thirty participants each year, only about 25 percent were from developing countries (infoDev 2000).

The U.S. Global Information Infrastructure and the ITU

In his bid for election to the U.S. presidency, Bill Clinton (1992) gained the support of high technology industries with the idea of a new infrastructure based on information. The Global Information Infrastructure (GII) envisaged a fully liberalized series of private global networks linked together by regulation governing interconnection. The GII provided a vision of a technologically determined future: a reinforcement of the supremacy of private industry over nation-states. Vice President Al Gore became the public voice for this technological utopia—to be also rapidly adopted by both Japan and the European Union in the interests of their equipment industries (Finnie 1995).

The ITU-D's first plenary in Buenos Aires in 1994 presented an opportunity to bring the message of the GII to the developing world. It came during a particularly bad period for both developing countries and for ITU-D.[12] Least-developed countries had overall debt liabilities as calculated by the United Nations Conference on Trade and Development (UNCTAD) of US$112.6 billion in 1990, compared to US$53.6 billion in 1982. Their accumulated external debt represented more than 60 percent of their combined GDP.[13] Impoverishment had increased their numbers in ITU membership from twenty-five in 1973 to forty-seven in 1994. Of the twenty-nine least-developed countries in Africa, only ten had more than fifteen thousand subscribers.[14] In other words, the national network of these countries contained fewer customers than a small town in the industrialized world. They needed between $50 billion and $90 billion in telecommunications investment, whereas the 1989 plenipotentiary had set aside only $2 million per year to help them. At the same time, cuts in overall ITU spending had left ITU-D struggling.

Why, then, did Al Gore bring the GII to the ITU and developing countries? One possible answer is that the GII was designed to aid progress in the General Agreement in Trade and Services negotiations, where few developing countries were taking part in the Negotiating Group on Basic Communications (see chap. 6). Another explanation is that the U.S. GMPCS operators were reliant on developing-country markets. Also following the Gulf War of 1990, the United States was unwelcome in the Middle East and needed new equipment-export markets. The total African market alone was worth $1.5 billion annually and growing fast. But Africa was the traditional province of the European companies, such as French Alcatel, German Siemens, and British Cable and Wireless. AT&T and others were trying to enter the market, and the ITU provided a forum.

The U.S. initiative was rewarded in ideological terms. The Buenos Aires Declaration incorporated five principles that the United States considered to be a "keystone for development of the global infrastructure": private sector investment; competition; a flexible regulatory framework; open access to all information providers; and universal service.[15] Here was the formal acceptance of the role of private companies in development: fruition to the Hansen Report of 1989. The ITU-D's members could now be expected to welcome U.S. private initiatives.

Other concrete benefits followed. In 1993, prior to the election of the African National Congress (ANC) government in South Africa, Alcatel had inaugurated a new optic fiber cable direct to Cape Town from Europe, ignoring the rest of Africa. Yet, because of the continent's reliance on old colonial networks, each year Africans spent more than $250 million to transit Europe in order to communicate with other Africans. In 1993 the ITU requested AT&T to investigate the possibility of providing Africa with its own submarine cable.

Critics argue that the ITU did not seek competitive bids for this project or consider satellite technology (Egger and Fullsack 2003). The ensuing project, "Africa One," to be financed by World Bank loans to African governments, was to encircle Africa with 39,000 km of fiber-optic cable connecting with forty-one African states and was worth $2.65 billion to AT&T (AT&T 1995). But the project held unfortunate connotations with the nineteenth-century encirclement of Africa by the British Eastern Company. With little influence in the operation and upkeep of the technology, and with already existing plans for expensive contributions to a regional African satellite (RASCOM), it was unclear how far the African continent would benefit. NGO critics, such as Probe International, argued that because World Bank intervention in Africa was a failure, it was now concentrating on building up infrastructure so as to propel market reforms (cited in Majtenyi and Fleet 1996).[16]

That a number of members were not happy with the new U.S./World Bank-supported ITU message of liberalization is evident from the Green Paper for Africa published in 1996. Although first drafted in 1991, the Green Paper took five years to agree on and ended only as "a thought-provoking reference document," not the blueprint for liberalization first intended (ITU

1996, vii). The eventual Green Paper for Africa reflected both the reluctance of African governments to move away from their traditional support for state ownership and the influence wielded by the ANC, elected to South African government in 1994. Rather than the neoliberal model of development that was still the mainstream World Bank's view, the report includes a paper on models of regulation written by World Bank telecommunications staff (ITU 1996, 149). Throughout the 1990s the ITU provided a communications forum for the World Bank telecommunications staff to publicize their views to developing countries and perhaps to leverage their ideas into the World Bank's mainstream (see chap. 5).

Restructuring: The Rise of the Private Sector

The Kyoto plenipotentiary of September 1994 was the first under the new rules of the ITU, and the first to adopt a strategic plan for the organization, so that it would become "the international focal point" for telecommunications (Maclean 1995). It was agreed that the ITU required enhanced private-sector participation and needed to broaden the scope of its activities from the technological to regulatory and policy matters. It also required partnerships and alliances with other international organizations.

Of particular concern was the role of industry. To some extent, by cooperating with other international standards bodies, the ITU-T sector had overcome much of the previous overlap in standardization. In 1996 it was said to have set 85 percent of all current international standards (Shurmer and David 1996, 4). But, with standards often determining market success, manufacturing industry wanted a greater role (Bellchambers 1994, 34). The ITU was faced with the conundrum of an intergovernmental organization needing its government members to cede power to private industry. The institution also needed to establish global legitimacy and as part of its rethinking began to consider for the first time the contribution that NGOs from civil society might make (Ó Siochrú 1995).

But Tarjanne's proposals to give more influence to the private sector in the standards process came up against opposition from the United States and its determination to define and control "U.S." industry (*ITU News*, 6/97: 5). Having previously pushed for more power for private industry, the altered U.S. attitude arose from the 1997 WARC. Because U.S. policy had centered on gaining near-exclusive frequencies for the satellite operator Teledesic Corporation (partly owned by Bill Gates) in the Ka-frequency band, a U.S. competitor, Celestri LEO, owned by Motorola, had turned to France to gain support. Just as the U.S. administration had responded to Western Union's 1930s actions by ensuring from 1947 that companies had to apply to national governments for permission to attend the CCIs, so in 1998, despite the internationalization of ownership, it opposed loosening such government control.

The European Public Telecommunications Network Operators Association threatened to withdraw from ITU T in an attempt to further reform. Some companies, such as BT, argued that the intention of the U.S. government was to see ITU influence in standards pass to U.S.-based organizations. But the U.S. State Department countered, "We are not convinced that drastic reform is needed" (Moloney 1998b, 1 and 27).

Pressure on the ITU mounted as the private sector, now contributing 13 percent of the budget, complained that its membership fees were five times those of member states. In 1998, Rutkowski recounted, "When I left the ITU and went back into private industry, I'll never forget the emphatic statement by a leading industry corporate president in a planning meeting—'We the industry are the ITU today'" (cited in *Communications Week International,* October 5). In 1999 the ITU was forced once more to review the rights and obligations of member states and sector members (Res.PLEN/11 *ITU News,* 1/99: 16). It was once more back to examining the rationale for its existence.

Global Mobile Personal Communications Systems

During this ongoing debate on the role of the private sector and civil society within the ITU, the problem of the U.S.-based Global Mobile Personal Communications Satellites arose. Conceived in 1985 and evolved from military technology, GMPCS operators used low and medium earth-orbiting satellites to provide mobile communications. Being closer to the earth (four hundred kilometers to ten thousand kilometers) than geostationary satellites (thirty-six thousand kilometers) these satellites were subject to less potential disruption of the signal and allowed ground stations to be handheld.

GMPCS were not providers of international facilities in the traditional sense but were end-to-end systems replacing national and international infrastructure. They raised issues of licensing, not only of the space segment but also within national systems. Despite British opposition to FCC unilateral action, in 1995 the FCC licensed Motorola Satellite Communications' Iridium as the first of five potential systems.[17] Following the FCC action, the British DTI then licensed ICO—a privatized offshoot of Inmarsat. These unilateral actions already bypassed the ITU, which was unsure how to approach the regulation of the new operators (Kelly 1994, 39).

The problem for the GMPCS operators was that, in the time taken to develop the technology, the market for global mobile systems had been filled with second-generation cellular phones and the spread of the European GSM standard. By 1995, GSM systems had already been adopted in seventy-seven countries and had 7.8 million subscribers (Mullins 1995). Although the satellite systems were handicapped by the potential cost of handsets at between $700 and $3,000 each and the price of calls—from $1 to $3 per minute—they needed to enter markets where GSM had not taken hold (Nelson 1998). To do so meant negotiating with each administration separately—both costly and time consuming (Mondale 1994). Following the failure of the WTO negotiations to open up the markets of the major developing countries (see chap. 6), they turned to the ITU in order to establish a dialogue. As in the GII debate, the ITU was useful to the U.S. government and industry as a gatekeeper to the developing countries.

In November 1994 the ITU brought the two sides together (Tyler 1995). The result was a memorandum of understanding (MOU), between the operators and governments, formally adopted at the first World Telecommunications Policy Forum of 1996. Those governments signing the MOU agreed to mutual recognition of equipment and a simplified licensing process for handheld terminals. But a number of developing countries, including Bangladesh, China, India, and Pakistan, withheld support (Moloney 1996, 1). Because of their objections, the final document was amended to allow individual nations to monitor satellite traffic within their borders. But the concessions were not enough to reassure them that GMPCS would not bypass their networks or add to their security problems. In other words, like Comsat before them, U.S. satellite operators found that they could not simply override the rights of states.

For the ITU this was the first of a series of MOUs that it facilitated (see discussion on ICANN on p. 138), but it was unclear how these were to fit into the organization's rationale. That rationale had already been called into question in 1994, when the WTO was founded and the United States and the European Union had proposed a WTO Telecommunications Committee to replace the ITU—a committee that would give the European Union voting rights, unlike the ITU. The intention was that such a committee would review the implementation of the WTO Agreement, formulate proposals, consider amendments to the Telecommunications Annex, provide technical assistance to those developing countries negotiating accession, and cooperate with international organizations active in the sector (Pipe 1993, 30). Under the proposals the ITU would "continue to regulate how governments manage the frequency spectrum, provide a forum for the development of technical standards and, through its technical cooperation activities, assist on the development of national telecommunications infrastructure in many of its member countries" (Tritt 1992, 16). However, the central role in telecommunications regulation would pass to the WTO committee that would oversee the liberalization of national markets.

But U.S. delay in accepting the proposed trade organization and its desire to keep the WTO small prevented the proposal from coming to fruition. In fact, ITU personnel took part in the WTO telecommunications negotiations, including the Telecommunications Annex 1992–94, the Negotiating Group on the Basic Agreement on Telecommunications 1994–96, and the Group on Basic Telecommunications 1996–97. In particular, having been responsible with the World Bank for opening up debate on national, independent regulation, the ITU contributed to the drafting of the reference paper on regulation that became part of the 1997 agreement (see chap. 6).

In 1997 Tarjanne saw the role of the ITU (in relation to the WTO agreement) in terms of information gathering, analysis, and dissemination among its 188 member states and over four hundred sector members. But he also wanted a formal alliance between the two organizations and observer status at the WTO's Services Council (Tarjanne 1997). Cooperation with the WTO was cemented at the ITU World Telecom Policy Forum in 1998 when the ITU agreed "to support member states in establishing regulatory bodies independent from telecommunications operators" (Moloney 1998a). A week later, the ITU announced that it would finance liberalization training centers in regions where the infrastructure of competition was least well developed. On May 26, 2000, the two organizations made a formal agreement (Mathew 2003, 207–15).

The WTO gave the organization a new focus—national governments in the developing world. In fact, led by the South African regulatory agency, perhaps seeking to expand its influence, the ITU tried to persuade developing countries to set up regulators, not only independent of the operator, but also independent from the prevailing national ministerial structure (Hills 1998a). The ITU found a role in pressing developing countries to liberalize and regulate. In turn, the emphasis on regulation gave the FCC more influence within the organization. The ITU held what was to be the first annual meeting for regulators from developed and developing countries in 1997, with the FCC in the chair. Hence the ITU had gained a role as the FCC and WTO's training institute and ITU-D had moved center stage.

Bypassing the ITU—"Accounting Rates"

Even while the United States was utilizing the ITU as a gatekeeper to the developing countries and refusing to cede power within the organization to industry, it was also attempting to break the ITU's control over the system of "accounting rates" that supported the state-to-state organization of international communications. The issue was not new, but it was brought to a head by overall deterioration in the U.S. balance of payments during the 1980s, of which, by 1989, about $1.4 billion (about 1 percent of the total) was contributed by the deficit on telecommunications services.

The system of state-to-state bilateral agreement on settlements with its 50/50 split was suitable to a system where there was near equity in outgoing traffic. But problems arose because European PTOs used profits from high international tariffs to cross-subsidize local calls, thereby increasing outgoing traffic from countries, such as the United States, where collection charges were lower. The CCITT attempted to remedy a growing disparity between settlement rates and collection charges in the 1960s without effect.[18] European PTOs continued to use international retail tariffs (collection charges) to cross-subsidize the expansion of the local network.

The ending of the Bretton Woods fixed exchange rates in 1973 forced administrations to "agree definitions within the CCITT which clearly distinguished between collection charges and accounting rates." "Accounting rates" were supposed to be linked to costs of transmission and, for European administrations, the CCITT developed a method of cost identification for individual components of the international network—switching and transmission.[19] However, a survey of "accounting rates" among OECD countries in 1988 found that although the trend was downward, even within Europe there were disparities of 50 percent or greater (OECD 1994, 69). With the decline in costs of infrastructure, the rates increasingly diverged from costs.

Because a country with an excess over 50 percent of outgoing traffic had to pay its bilateral partner, those countries with more outgoing than incoming traffic had an incentive to negotiate low "accounting rates." Those with more incoming than outgoing traffic wanted high "accounting rates." Both European and developing-country operators relied on settlement payments for the hard currency they brought into the national coffers.

Looking for ways to fund the expansion of systems in developing countries, in its 1985 report the Maitland Commission recommended a rearrangement of "international traffic accounting procedures with the aim of setting aside a small proportion of revenues from calls between developing countries and industrialised countries" that could then be put to developing infrastructure in poor countries (ITU 1985, 9, para. 30). It pointed as a potential model to the favorable 60/40 split on accounting settlements adopted within the British Commonwealth.

The Commonwealth Model

In fact, the ITU system of bilateral "accounting rates" had only been adopted in the British Commonwealth in the 1980s. Prior to that time it had used a "wayleave" scheme. Under this scheme, initiated after the nationalization of Cable & Wireless in 1949, the sending country kept the collection fee it charged and then shared the costs of the upkeep of the telegraph cables in proportion to that revenue.[20] So, for instance, if the Post Office received 40 percent of the total Commonwealth revenue, it would pay 40 percent of the common user expenses minus whatever costs it was allowed to claim. The differences would then be settled as net wayleave payments or receipts, as the case might be (Commonwealth Telecommunications Bureau n.d., 5). The system was cheap to run and originally benefited the British by contributions to the capital costs of Cable & Wireless's network.

But the very profitable transatlantic traffic with the United States via the TAT cables threatened to make the British contribution too high. As a result, in 1959 a second wayleave scheme was created solely for telephone, telex, and leased lines over coaxial cables. The basis of settlements was also changed to a fixed scale of "accounting rates," varying according to distance, instead of actual receipts. But the British refused to include traffic generated by Intelsat in the scheme so as to limit the British "subsidy."[21]

As increasing numbers of British colonies became independent, a permanent organization was created in 1967—the Commonwealth Telecommunications Organisation—part of whose function was to oversee the wayleave scheme. Then, in the early 1960s, because India felt that it was paying too much, it was allowed to freeze wayleave payments and the United Kingdom paid out an additional £200,000 for three years.[22] By 1970 the British Post Office estimated it was paying out annually £1.5 million too much into the scheme and determined to move the Commonwealth to the ITU system of bilateral "accounting rates."

British entry into the European Union in 1973 provided it with the necessary excuse because the wayleave scheme contravened the nondiscrimination clauses of the Treaty of Rome. But the political timing was inopportune. The alternative system introduced in 1973 was designed to limit the amount paid by outlying, poorer members to their actual usage of specific parts of the network, but it ended up benefiting primarily Canada and Australia (HMSO 1973). The system foundered on what constituted "costs" and lack of central power to regulate national administrations.

Finally, following the introduction of competition into the British market and Mercury's opposition to subsidizing other private companies in the commonwealth, in 1983 the commonwealth financial agreement was replaced by a variant of the ITU's bilateral system. BT, together with the incumbent operators in Canada, Australia, and New Zealand, agreed to a 60/40 split in settlements with developing countries within the commonwealth up to a specified amount. But gradually these donor countries preferred to allocate funds to training staff of developing countries in digital technology, often at Cable & Wireless's own training facilities. As a result, in 1990, in letters between governments, it was agreed that the commonwealth preferential "accounting rates" arrangements should end. Hence, the commonwealth system was already dying as the Maitland Commission recommended it.

The U.S. Backlash

Perhaps because of the commonwealth system's demise, at WATTC-88 and at the 1989 Nice plenipotentiary developing countries pursued the Maitland proposal for a 60/40 percent sharing of settlements. But European PTOs benefited from an excess of traffic with the United States so that in 1987 BT received $57 million, Deutsche Telecom $167 million, and Italy $51 million in settlements (Stanley 1988). CEPT had no wish for "accounting rates" to reach the agenda. In an attempt at WATTC-88 to bury the issue (admitted privately), all CEPT members, with one exception, now signed up to demand a comparative study of "costs" with which they had previously refused to cooperate (ITU 1988, Resolution PL/3). This study eventually showed that, on average, developing countries faced "costs" 2.08 times higher than those of carriers such as AT&T—and in some cases 4.1 times higher.[23]

Meanwhile, following the breakup of AT&T and the entry of MCI and Sprint into the international market, in 1986 the FCC made its International Settlements Policy applicable to telephony. Under this policy the FCC mandated an equal division of "accounting rates" between foreign and U.S. carriers; the nondiscriminatory treatment of U.S. carriers; and a proportionate return of inbound traffic—in other words it expanded its 1943 international formula from telegraph to telephone. The result was competition for market share of outgoing traffic so as to obtain the proportionate return of incoming traffic. Operators increased country-direct and callback services, which, for settlement and billing purposes, treated calls originating in foreign countries as if they had originated in the United States. The result was a burgeoning U.S. settlements deficit.

The FCC chairman, Dennis Patrick, for the first time spoke out about the issue in February 1988, and in 1989 "accounting rates" presented the opportunity of a win-win issue for new chairman Fred Sikes. Action could appease a protectionist Congress. The FCC's first report on the matter found that the U.S. deficit for international services was $1.7 billion in 1987. In total the United States had deficits with 171 countries, of which Mexico's ($313 million) was the highest, and surpluses with seventeen countries, of which Canada's ($19 million) was the highest.

According to the FCC, factors causing the disparity of outbound to inbound traffic included competition among U.S. carriers, high foreign tariffs, high international "accounting rates," and a weak dollar. So, for instance, a 10–minute call from the United States to Italy cost $11.75 in 1988, whereas the same call from Italy to the United States cost $30.70. The FCC argued that in some cases, such as West Germany and Italy, AT&T was charging its customers less than the "accounting rate" charged by the corresponding administration.[24]

By drawing further attention to the issue in 1988 and 1989, the developing countries helped to provoke a backlash from the United States, particularly when Iraq demanded a 60/40 settlement split with AT&T (*FCC Week*, April 17, 1989). Backed by Australia, the United States took the issue to the OECD, which found disparities between "costs" and charges to customers for calls to the United States of between 10 and 200 percent (cited in *Transnational Data and Communications Report* 1990, 6).

After an interview with Sikes, in April 1990 the British *Financial Times* began a campaign against what it called the "international cartel" that was keeping international call prices high (Dixon 1990a, 1990b). The campaign had immediate effect in the United Kingdom, where Oftel announced that it would cap international call rates (Dixon 1990c). Oftel also began to publish British "accounting rates" in order to influence the debate and, reacting to government pressure, U.S. and British operators agreed to halve them by 1993 (Dixon 1991a).

In May 1991 the FCC directed U.S. telephone companies to negotiate lower "accounting rates," particularly in Europe and Asia. It threatened further unilateral action if there had been insufficient progress by 1993 (Dixon 1991b). By the time that the CCITT Study Group III considered the matter in September

1991, Australia, Canada, Sweden, New Zealand, the European Commission, and the OECD had joined the United States in its campaign. Sweden had previously proposed that sender-keep-all arrangements should be adopted within CEPT. However, the CCITT Study Group defeated the radical proposal from the OECD for the introduction of a "termination charge" that would have meant an identical charge for local and international interconnection. In addition, it rejected publication of rates and the U.S. demand that international tariffs should take account only of the costs of international transmission. These "cost-based" tariffs would have favored international operators and deprived developing countries of including the capital costs of their national networks in the settlement rates (Schenker and Lynch 1991). This was the first indication that "costs" were to become an international political issue.

The following year, in 1992, the new ITU-T Study Group III reached agreement that "accounting rates" should be "cost-based, non-discriminatory and transparent" (Recommendation D-140).[25] However, the concept of "cost-based" was considerably different from that proposed by the United States. The ITU construction of costs included three elements—international transmission facilities, international switching facilities, and the national extension—each of which included both direct and indirect costs, such as universal service.

In 1992 the FCC chairmanship changed under the incoming Clinton administration to Reed Hundt, perhaps more concerned with international matters than any previous chairman since Fly. He was characterized by his predecessor James H. Quello as "Attila the Hundt" and castigated for the extent to which he "and the executive branch tried to exert their influence on an independent agency created as an arm of Congress" (Quello 2001, 93). Under Hundt's chairmanship, supported by the USTR, the FCC proposed benchmark settlement rates of between 23 cents and 39 cents to terminate U.S. telephone calls in Europe and between 39 cents to 60 cents to terminate U.S. telephone calls in Asia and other regions (FCC 1992). Unless "accounting rates" came down, it proposed to impose the benchmarks in 1993.

Observers linked this unilateral declaration with the FCC goal of worldwide international "resale" and the fact that few countries had agreed to it. Writing in 1993, Yukio Ito suggested that "the FCC is tacitly trying to force the world to choose between two alternatives . . . either accept participation by U.S. carriers in the resale market or accept a reduction of the international accounting rate" (1993, 14). Other commentators linked the FCC action to AT&T's interests. On this understanding, AT&T's original position was to support the liberalization of international "resale." An increase in resellers would bypass accounting rates and exert pressure to bring them down— something for which AT&T had traditionally argued.

However, having first pushed for "resale," AT&T changed position in 1992 and lobbied the FCC to regulate the service. Facing declining profits and massive layoffs, at first "resale" may have seemed to provide AT&T not only with lower accounting rates but with the opportunity to fully utilize its fiber-optic cables. But then, after second thoughts, "resale" threatened an increase in its overseas settlement payments. Because of high public network tariffs, overseas callers were more likely to use resellers and evade the "accounting rates," whereas low tariffs in the United States meant callers were more likely to use the public network (Ito 1993, 13). Subsequently both AT&T and MCI established "resale" subsidiaries in Japan and Europe in order not to lose customers, but AT&T pressure led to FCC regulation of international "resale" on a reciprocal market entry basis, thereby slowing its development.

In the meantime the unilateral benchmark decision by the FCC caused international outrage. The FCC had deliberately bypassed the ITU. For the first time 1993 figures showed that much of AT&T's deficit with European operators was accounted for by country-direct services directly attributable to the FCC's International Settlements Policy (Aamoth 1994, 28–29). In the light of these figures, and the fact that the settlements deficit declined in 1992, the basis of the AT&T and FCC case changed from the size of the U.S. deficit to the fact that "accounting rates" were above "costs."

Challenged that it had no data on "costs" on which to base its benchmarks, the FCC retreated for a while. However, AT&T continued to put pressure on developing countries with "name and shame lists" of those it considered had the worst accounting rate practices. U.S. negotiators then changed the forum for the issue to one where developing countries had less power.

The U.S. delegation first introduced the issue into the WTO negotiations in 1992, but Brazil and India in alliance with the European Union prevented it from including price issues in the 1994 Telecommunications Annex. The United States then reintroduced the issue of "cost-based" interconnection tariffs into the WTO's Negotiating Group on Basic Telecommunications and (as chap. 6 explains) was eventually successful. By forum-swapping to the WTO, the FCC bypassed the ITU.

ICANN: Bypassing the ITU

The U.S. government once again bypassed the ITU in the regulation of the Internet, preferring to set up a private corporation under U.S. government control. Again, the issue has resonance with that of Comsat—a U.S. monopoly of a new technology funded by U.S. defense agencies that led to global unilateral regulation. Similarly, although the matter is still ongoing, the loss of legitimacy of that unilateral solution has led to the search for an intergovernmental alternative.

The Internet itself evolved out of a project by the U.S. Department of Defense to decentralize computers in order to protect them from nuclear attack and began with the innovation of a Domain Name Server (DNS) in 1984 that converted the numerical addresses of computers to an alphabetical system and vice versa. Created first as Advanced Research Projects Agency Network (ARPANET), purely a defense-related packet-switched data network, under the aegis of the U.S. government's National Science Foundation it then became NSFNET, run by a combination of volunteers and U.S. and civilian contractors. It was a U.S.-government-funded, government-controlled data network, and as late as 1993 it was primarily a North American network.

By the late 1990s on both the technical and regulatory front the ITU had come under competition from the Internet. On the technical front the Internet's open TCP/IP standard for packet switching was replacing the ITU's telecommunications standards. Telecommunications operators in the industrialized countries were beginning to consider migrating their networks to the IP standard. But the unreliable technical standards of the Internet were determined by two hundred or three hundred individual engineers involved in its creation, the Internet Engineering Task Force, not the ITU.

Similarly control of the Internet's top-level domain name system (.com, .org, .net) rested with a group termed the Internet Assigned Numbers Authority (IANA) contracted to the U.S. government, of which Jon Postel, an academic, was head. The registration of users onto the A-root server (the controlling core server of the Internet) was done by a private company, Network Solutions Inc., under a monopoly license from the U.S. National Science Foundation, set to expire in 1998. The U.S. government and the company were in conflict over ownership of the registration data.

Further problems arose because the net was conceived as a cooperative, academic means of exchanging information, not a commercial entity. By 1995, commercialism meant popular domain name addresses such as .com were running out (Hart 1996a, 7). Other problems ranged from lack of security and reliability, to the settlements system that European Internet service providers regarded as unfair. Because the most popular Web content and the most network access points for interconnection were in the United States, European providers found themselves paying the cost of a full transatlantic circuit (not the half circuit of the state-to-state model). Europeans argued that they paid

80 percent of the transmission costs of customer access to the Internet and that the model was "U.S.-centric" (cited in Mendler and Hart 1996, 3). In turn, these easy profits led to an expansion of U.S. international carriers.

In 1996, Postel put forward a plan for Internet expansion under which fifty new domain name registries would each be responsible for administering three top-level domain names (Hart 1996b, 3). Although opposed by the Internet Engineering Task Force as commercialization of the Internet, a group of individuals from the Internet Society, the World Intellectual Property Organization, the International Trademark Association, the IANA, and the ITU formed themselves into an international ad hoc committee to examine and propose changes to the domain-name system.

The ad hoc committee released a study on generic top-level domain (gTLD) management that resulted in agreement in 1997 of a generic top-level domain memorandum of understanding (gTLD-MOU) for which the ITU agreed to act as a depositary for the signatories (ITU 2002, 2.1). This MOU proposed assigning governance functions of one top-level domain name (.int) to an entity housed in the ITU, with representation from business interests, international governmental organizations, and the Internet Society—but not Network Solutions Inc. or the European Union (Hart 1996a). The ad hoc committee then convened a meeting in May 1997, to be hosted by the ITU, where its proposals for seven new top-level domain names would be discussed and the MOU signed.

However, in the United States, former ITU official Anthony Rutkowski argued that the MOU amounted to an intergovernmental agreement initiated by a nonlegal entity. The European Commission also objected because it had not had representation on the ad hoc committee, and U.S. Secretary of State Madeline Albright claimed that the ITU was entering into an international agreement without the consent of member states. Network Solutions Inc., seeing its monopoly challenged, lobbied Congress. And further criticism came from the registries of country-level domain names (the country-code top-level domains [ccTLDs] such as .uk), which had not been included (Kleinwächter 2004). Protests from around the world accused the "gTLD gang" of using despotic tactics to profit personally from the Internet (Loundy 1997). Others contended that the hierarchical PTOs of the ITU would take over the nonhierarchical Internet. Tarjanne argued that the MOU signed by fifty-seven entities was an example of a new paradigm of "voluntary multi-lateralism" (cited in ITU 1997a), but such was the furor that the ITU and the World Intellectual Property Organization (WIPO) did not sign.

In July 1997, refusing to accept Postel's plan to bring the A-root server under the control of the ITU, the White House issued a presidential directive along with *The Framework for Global Internet Commerce*, unofficially attributed to Al Gore. The framework mapped out the Clinton administrations' vision for a "non-regulatory, market-oriented approach" to the digital marketplace (U.S. White House 1998). One critic called it "NAFTA redux in cyberspace" (cited in Freed n.d.). Then followed a U.S. green paper in January 1998, which proposed that the U.S. Department of Commerce should take the position of a licensing authority over fully commercialized registries (ITU 2002, 2.2).

Overseas governments warned against the U.S. government's proceeding unilaterally. In 1997, OECD governments had met for the first time to consider the issue of domain-name registration, and in 1998 the European Union issued a press statement criticizing U.S. dominance over the Internet (Cukier 1997, 7). The European Union was concerned that WIPO and the ITU had been disregarded in the green paper and wanted more involvement of E.U. private industry. Following a joint proposal from Commissioner Martin Bangemann of the Information Industries Directorate and Sir Leon Brittan of the Competition Directorate, the European Union called for an international representative body for future Internet governance (Kleinwächter 2003, 1111). Postel took action, demonstrating the potential he held for disruption of the Internet, and the green paper died.

Following adverse comment on the green paper, the U.S. Department of Commerce issued a statement of policy in June 1998, known as the "White Paper." This paper defined broad principles and procedures that the U.S. government would use to transition from its existing management role to a "new nonprofit corporation." But still the U.S. administration opposed a role for governments in the management of Internet names and addresses (ITU 2002, 2.2). The white paper called for oversight of Internet names and addresses by a new, private, not-for-profit corporation headquartered in the United States. The U.S. government would maintain an oversight role. Under the leadership of Ira Magaziner, adviser to President Clinton, the intention was to privatize the Internet and introduce competition in the .com registration market. However, the white paper was perceived by critics as presenting a more authoritarian model than the green paper (Freed n.d.).

Shortly after publication of the white paper, an international group incorporated ICANN as a private, nonprofit, California corporation (ICANN-watch n.d.). Postel drafted ICANN's bylaws and selected its initial board of directors. He preserved the corporation's independence from governments. Under ICANN's original bylaws (ICANN 1998b, Art. V, Sect. 5) national government and international entity officials were ineligible to join the board of directors. But Postel had taken into account the demands of the European Union for involvement in the regulation of the Internet. He now included a governmental advisory committee (Art. VII) to "consider and provide advice on the activities of the Corporation as they relate to concerns of governments, particularly matters where there may be an interaction between the Corporation's policies and various laws, and international agreements." But the committee's recommendations were nonbinding on the board, which had only to consider governmental advice on those proposals on which it sought comments (ICANN 1998b). In other words governments were to be subordinate to the apolitical board of individuals. The ITU was neither involved in nor aware of the fact that these discussions had taken place, nor of the "creation and particular legal formulation of GAC" (ITU 2002, 2.5).

Postel died suddenly on October 16, 1998, but had already persuaded the U.S. Department of Commerce to transfer the contract for IP addressing from IANA to ICANN. Following his death the department delayed while Ira C. Magaziner phoned around to see if ICANN was acceptable without Postel's backing. Magaziner said that the Clinton Administration's consent to the ICANN proposal would hinge on its supporters reaching an agreement with those calling for a more open organization (Harmon 1998). The Department of Commerce would retain its oversight role "until such time as the new corporation was established and stable, phasing out as soon as possible, but in no event later than September 30, 2000" (ITU 2002, 2.4). In fact, it was then extended several times.

Having previously said that he considered the ITU to be the place for regulation of Internet technology, on the eve of his retirement in 1998, Tarjanne had to reassure the Clinton administration that the ITU was not attempting a takeover of Internet regulation—that "it had no desire to grab any turf or target or control the Internet" (*Computergram International* 1998). Wolfgang Kleinwächter points out (2004, 239) that the gTLD-MOU was ignored at the ITU plenipotentiary conference in Minneapolis in October–November 1998. Even the ITU Resolution 102, which referred to "Management of Domain Names and Internet Addresses," did not refer to the gTLD-MOU but, instead, invited the secretary general of the ITU "to take an active part in the international discussion and initiatives of the management of domain names and Internet addresses, *which is currently being led by the private sector*" (emphasis added). The resolution stated inter alia that "the methods of allocation of Internet domain names and addresses should not privilege any country or region of the world to the detriment of others" and that "governments should promote a fair and competitive environment among companies or organizations responsible for Internet resource allocation" (cited in *ITU News* 10/98, 17–18). But there were those such as Rutkowski (1998, 8), who argued that attempts by governments to exercise sovereignty over the Internet were "neither essential nor appropriate."

Kleinwächter (2004, 239) argues that a "Minneapolis Deal" was reached during the ITU plenipotentiary in classic diplomatic "horse trading." The idea of a world conference on the information society had come out of UNESCO, but the U.S. government was not prepared to have that organization host it and had previously also opposed the idea of the ITU as host. But at Minneapolis the "U.S. government withdrew its opposition to the plans of the ITU . . . in exchange for the recognition of the private sector leadership in internet governance."

Eight days later, the interim ICANN board of directors had its first meeting in Cambridge, Massachusetts. And on November, 25, 1998, following a letter from Commissioner Bangemann stating that the European Union approved, the Department of Commerce recognized ICANN as the "NewCo" (ICANN 1998a). The memorandum of understanding between the Department of Commerce and ICANN was for a transitional period only. After two years and under the condition that ICANN fulfill its functions, the department would transfer the remaining responsibilities, including complete control over the A-root server, to ICANN (Kleinwächter 2004).

Despite the creation of a domain-names supporting organization by ICANN in the spring of 1999 to make recommendations on how ICANN should oversee the domain-name system, opposition to the lack of democracy in the organization, to its secrecy and lack of accountability continued, not least in Congress.[26] An attempt by the interim board to redefine itself as the "initial" board and extend its term of office did nothing to inspire confidence. Some critics claimed that it had been captured by the same "privateers" who attempted "to usurp control of the Internet in 1997, the same old bandits behind new masks" (cited in Freed n.d.).

Attempts at elections to the directorate from an at-large council of self-nominated individuals did little to fend off criticism that ICANN was accountable to no one but the attorney general of California, where ICANN was incorporated. ICANN's subsequent institution of mandatory arbitration of trademarks—its uniform dispute resolution policy—made evident its pivotal role as a global regulator and brought claims from Karl Auerbach, an elected director of ICANN, that ICANN was enacting a worldwide law "distinct and different from any enacted by any national legislature." He suggested to Congress, "The real question is not whether ICANN ought to have this power [supranational scope] but rather how ICANN's possession of that power obtains acceptance and legitimacy from the nations of the world and the users of the Internet" (Auerbach 2001). Questions of interconnection and legitimacy brought the ITU back in.

Increasingly, telecommunications services and the Internet had to interconnect so that in 1999 the ITU signed an MOU with ICANN to become a founding member of the ICANN protocol supporting organization and collaborate on IP network standardization issues (ITU 1999b). Criticism of ICANN from governments gave the ITU additional leverage. In 2002 the ITU secretariat wrote, "The ITU Secretariat considers that ICANN cannot be characterized as a 'private sector body' if its existence, particular responsibilities and oversight is subject to ongoing oversight of a single government." Instead, if a private/public sector partnership was essential, then "it must flow from full international cooperation" (ITU 2002, annex A, 3b). In other words, rather than accept what the Clinton administration would have the world believe—that ICANN was Internet "self-governance"—the ITU and others interpreted the organization as directly subordinate to U.S. power.

From 2002, as developing countries found themselves increasingly marginalized in terms of the "digital divide," the issue of Internet governance became part of the multilateral discussion on the UN/ITU World Summit for an Information Society (WSIS). The ITU became both a conduit for governments to try to gain control and of civil society NGOs' attempts to democratize access. However, U.S. industry continued to argue that "international Internet public policy issues should be done on an ad hoc basis and NOT through the current intergovernmental structure of the ITU" (U.S. ITU Association 2003; emphasis in the original). Just as Comsat had previously lost legitimacy, so did ICANN, seen by many developing countries as an organization dominated by the U.S. government. But because China and others had done what many regarded as impossible and had instituted national regulation over their citizens' access to the Internet, fear of the illiberal intentions of national governments balanced demands for change. In 2005, following a strongly worded letter from Secretary of State Condoleeza Rice to U.K. Foreign Secretary Jack Straw, the European Union backed down from its campaign for more intergovernmental control (Johnson 2005). U.S. control over the A-root server was once more confirmed in the interests of "stability and security"—in other words, the "war on terror."[27]

However, the lack of legitimacy and trust in a U.S.-controlled organization held the long-term commercial danger of a fragmentation of the Internet into national constituencies. Then, following a series of controversies surrounding ICANN's failure to establish a domain designation of .xxx for pornography (which the European Commission said was due to the Bush administration's intervention) and its agreement with VeriSign to manage the .com registry in perpetuity (which Congress said was a monopoly license for price hikes), the U.S. government submitted to domestic and international pressure. In July 2006 the Department of Commerce indicated that it was prepared to transition its control over to ICANN—although not to relinquish control of the root server (McCarthy 2006). The department followed this announcement in a three-year MOU with ICANN by reducing its direct control and by demanding greater transparency from the organization to presage that organization's full independence in 2009 (OUT-LAW News 2006; Broache 2006). The expectation is that the Government Advisory Committee of ICANN will become the conduit for some form of international governmental control, but that despite calls from countries such as Brazil, U.S. domestic politics will ensure that ICANN will remain outside the U.N. system. Questions remain over the internationalization of control of the root server. Hence, although the Internet currently remains under U.S. government control, as in the case of Comsat previously, sustained international pressure from governments coupled with domestic criticism has forced the United States to back down from unilateralism.

Conclusion

In the immediate post–World War II world, the industrial countries' interests were in upgrading their domestic networks. Then, as they pushed forward with the introduction of electronic switching and the complex standardization that went with it, they were less than supportive to any proposal that ITU resources be allocated to developing countries whose captive markets provided exports for outdated equipment. Only as it became evident that international communications might be compromised by lack of training and expertise did the industrialized countries (except for the United States and United Kingdom) become less antagonistic.

As the developing countries gained in numbers, so their voting power in the organization increased. The tension between the traditional focus of the work of the ITU on the standardization concerns of its industrialized members and the development problems of the majority of its membership was increasingly reflected within ITU plenipotentiary meetings, where the developing countries held the numerical majority. U.S. delegations appear to have had problems adjusting to the demands for an equitable redistribution of resources, both within the organization and in relation to those resources the ITU controlled (such as the radio spectrum). The United States used direct threats over the withdrawal of funding, as well as indirect power through procedural maneuvers and the organization's tradition of unanimity and consensus to bypass votes. The OPEC oil price hikes of the 1970s represented the peak of the power of developing countries within the organization, resulting in an eventual restructuring in their favor.

From the early 1980s and the liberalization of the domestic markets of the United States, the United Kingdom, and Japan, the old purpose of the ITU—to protect the interests of monopoly PTOs—no longer held universal validity. During the 1980s the primary conflict within the organization was north vs. south, with the United States and the Soviet Union in the same coalition. Technological change exacerbated these tensions. Large users demanded a say in standard making, and Integrated Services Digital Network (ISDN) standardization involved not only technical but policy matters that reached into the national level. Reacting to the perceived threat to its sovereignty and to its service industry, the United States led pressure to withdraw from ITU standardization activities in favor of collaboration between the standards bodies within each of the triad of industrial nations (United States, Japan, and Europe). To a large extent it was this threat to its major activity, not the increased voting power of developing countries, that provoked a restructuring of the ITU as an organization.

In the debate that ensued on the role of the ITU, the World Bank became influential. The previous underlying consensus on the state's responsibility for telecommunications gave way among its secretariat to an acceptance of the neoliberal demand for the restructuring of markets. From the 1990s, aided by the institution's personnel, the industrialized countries reasserted themselves. Business became increasingly influential.

Restructured into three sections, in which the developing countries were to talk to themselves, the ITU's financial and personnel problems meant it could speak of the "digital divide" but do little about it. The organization lost its way. Its leaders embraced the notion of the supremacy of markets and the loss of state sovereignty, thereby becoming a mouthpiece of the "Washington consensus" and opening the ITU for U.S. business opportunities.

The ITU was rescued by World Bank telecommunications personnel with their strategic need for a body that could help foster regulation in developing countries. Hence in 1993 the ITU held its first forum on regulation—four years before the World Bank acknowledged that states had a role in the market. By 1994 it had already become clear that the organization required a new focus.

But the decision to attempt to bring the ITU back to the center of the international telecommunications stage by expanding its membership to the private sector created conflict with member states. Private operators, equipment manufacturers, and others demanded increased influence, but it was the United States who refused to cede power. To have allowed corporations direct entry to the ITU would have reduced the power of the FCC back to its pre–World War II days.

The demise of the Soviet Union, the ongoing recession in the industrialized West, overcapacity in equipment manufacturing, and the concept of the GII with its worldwide liberalization of markets all provided a rationale for private companies to pay more attention to developing and transitional countries. At certain moments the ITU was useful to the United States as in the promulgation of the liberalization of world markets proposed in the GII to reinforce negotiations in the WTO.

The ITU could also provide the venue for operators from the industrialized world to meet developing country PTOs as in the MOU on GMPCS. But, here as elsewhere, the ITU leadership was perceived as biased in favor of the U.S. operators, at first ignoring opposition from the Far East. In fact, the FCC had already bypassed the ITU in licensing the GMPCS operators unilaterally, intending that they should operate worldwide, end-to-end networks. It was only when the companies then found that they could not evade the state-to-state system and that individual bilateral negotiation with each country was too expensive that the ITU was called into play.

Similarly, together with the European Union, the United States attempted to replace the ITU with a committee of the WTO to oversee the implementation of WTO rules. When that failed, the ITU was drafted into the negotiations to help bring developing countries onside. Then, once the WTO Agreement on Basic Telecommunications had been signed in 1997, the ITU entered into a cooperation agreement and became the WTO's mouthpiece and training institute.

But the ITU still held regulatory power over the payments system that supported the state-to-state structure of international telecommunications. Although at first content to use its political muscle to talk down the level of bilateral "accounting rates," under the chairmanship of Reed Hundt the FCC set out to use direct power to override the ITU's CCITT. When that initiative encountered opposition within the ITU, the USTR/FCC removed "accounting rates" into the WTO discussions—a forum more favorable to U.S. interests.

Developed under defense procurement in the United States, the Internet—a packet-switched data network using open standards—was singled out by the U.S. government for special treatment and subsidy. Moving from a defense network to an academic network, the Internet developed around informal arrangements and among people concerned to retain its noncommercial, cooperative features. But the introduction of the World Wide Web and the ease of creating home pages brought commercial entry and thousands of nets to be interconnected. Overcrowding led to attempts to expand the number of domain names and create new registries to administer their allocation. But the prospect of the ITU in 1996 becoming one of those registries, or gaining control of the A-root server, met with a U.S. furor. The ITU was humiliated by the Clinton administration.

When proposals for direct U.S. government control brought international antagonism, in ICANN the U.S. government created a private corporation. Unlike Comsat, ICANN did not come about through legislation and did not come under the FCC. ICANN exists as a private corporation only because of its relations to the U.S. government—a private corporation as the proxy for direct U.S. governmental power. The reported United States secret tradeoff with the ITU leadership—that the ITU might host the World Summit on the Information Society (WSIS) in return for not opposing ICANN—backfired. The WSIS became the venue for developing-country and E.U. attempts to alter the regulation of the Internet. Despite initial resistance, the U.S. government has been obliged to cede power, to "privatize" ICANN, and to allow some form of internationalization.

Hence, in the early years of the twenty-first century the ITU remains as the venue for intergovernmental discussion and groupings. The United States has found it difficult to control a U.N. institution, based on one nation–one vote. Withdrawal was the first strategy in the face of developing-country demands—but U.S. companies needed the standards function of the ITU. Restructuring was a second strategy, isolating developing countries, but then their markets became important. A pullout of standards work into regional organizations was a third, stopped by the Japanese. The use of private industry to provide surrogate control was another. But an increasingly internationalized private industry proved not altogether amenable to direct U.S. control. The World Bank has helped provide the United States with indirect power in the ITU through ideas on privatization and regulation, but replacement of the ITU's indirect influence over developing countries with direct power through a WTO committee did not succeed. Weak as it has become, and despite the dominance of private industry, the ITU has had the legitimacy to act as a conduit for democratic defiance against the U.S. "Empire project." Hence, bypass has become a favored U.S. alternative strategy—as in the creation of ICANN under direct U.S. control and in "accounting rates" discussed in chapter 6.

5

The World Bank and Privatization

This chapter explores the argument that the World Bank became party to the U.S. goal of opening up foreign domestic telecommunications markets to U.S. capital. As recession in the West reduced bilateral aid, so developing countries relied on foreign investment. The collapse of the Soviet Union in 1990 created further competition for such investment. In line with the ideology of privatization of the Reagan, Bush, and Clinton administrations, the World Bank restructured foreign domestic markets to sell state assets to Western companies as a condition of aid.

The outright sale of PTOs to overseas investors only took place in cash-strapped developing countries and where there was an authoritarian government able to withstand political backlash. However, from the early 1990s, cellular technology allowed governments to placate the World Bank and urban consumers without domestic political reprisals. In general, it was Western, state-owned operators and U.S. companies who benefited from World Bank–financed investment. It spread recolonization.

From the 1980s the mainstream World Bank viewed the state in developing countries as an impediment to the market, and the International Finance Corporation was happy to effect privatization without regulation. But, following the British model, the bank's telecommunications personnel pressed governments to regulate the sector in parallel with privatization. Finally, as the WTO Basic Telecommunications Agreement institutionalized domestic regulation, in 1997 the World Bank itself altered its policy to allow a role for the state in the regulation of privatized sectors—regulation became linked to the bank's traditional support for elites and limited democracy.

Background

The International Bank for Reconstruction and Development (IBRD) was established in 1944 as part of the Bretton Woods institutions that conceived the IMF and the unsuccessful International Trade Organization. The World Bank was narrowly conceived in terms of postwar reconstruction problems in Western Europe, but its original role was undermined by U.S. grants under the Marshall Plan. At the insistence of Maynard Keynes, the Bretton Woods conference adopted the principle of limited government's paid-in subscriptions so that the private capital markets mainly supplied the World Bank's resources (Mikesell 1972).

The IBRD's subsequent struggle to gain credibility with Wall Street led to a low rate of lending on infrastructure projects put forward by newly independent governments. By 1950 it had disbursed only $100 million out of the $350 million it had authorized. To increase loans in 1956 the International Finance Corporation (IFC) was created to finance private sector investment.

The World Bank's loans to poor countries in the 1950s stemmed from Western desire to counteract the influence of the Soviet Union. But provided at normal interest rates, these loans created rising indebtedness. The possibility that the United Nations might set up a soft loan fund led in 1960 to the establishment of the International Development Association (IDA). The three institutions—IBRD, IFC, and IDA—became the World Bank Group, with the IFC working separately until the 1990s. The group acted as a bank, a development agency, and a development research institution (Gilbert et al. 1997). Its power came in its ability to propagandize and implement the West's changing conceptualizations of "development."

U.S. Power over the World Bank

As in the case of Intelsat, the World Bank Group's location in Washington, D.C., has lent itself to the view that the World Bank is a U.S. domestic institution (Kapur 2002, 66). Although the number of U.S. citizens in the World Bank has fallen, U.S. education of its top personnel is a factor in allegations of excessive U.S. influence (Ritzen 2005, 111). According to tradition, a European heads the IMF, and an American heads the World Bank. Advised by the U.S. Treasury Department, the U.S. president hires and fires the World Bank's president. A number of insiders have cited the U.S. Treasury–Wall Street complex as the most important influence on the World Bank's policies and have documented the friction that can result between the U.S. administration and the World Bank (Peet 2003, 207). For instance, the sacking in 1999 of Joseph Stiglitz (then the bank's chief economist) for his outspoken criticism of the bank was on the orders of U.S. Treasury Secretary Larry Summers (Ritzen 2005, 101).

The World Bank is formally governed by a board of executive directors with votes weighted according to shares of capital. The United States' original share of 35 percent (now 16.41 percent) has been insufficient to block proposals, but, as in the ITU, the tradition of consensus increases its power (Klutznick 1988). Recent commentators suggest that the influence of the executive board has declined. Instead, bank presidents deal directly with governments on political grounds. Others argue that the Europeans have preferred dominance by the United States to dominance by one of their own. Yet, since 1991, the World Bank has faced competition from the European Bank for Reconstruction and Development (EBRD) created under European control (Kapur 2002, 59–60; Ritzen 2005, 103).

Congressional power over the World Bank increases when additional capital for the IDA is needed, and it has delayed U.S. contributions on numerous occasions (Gwin 1997). During the 1970s and 1980s, Congress threatened to legislate how IDA capital should be spent—something unacceptable under the bank's charter. The threats produced incidents, such as that in 1980 when Robert McNamara, World Bank president at the time, effectively promised not to make loans to Vietnam. At times, Congress has also legislated to control the vote of the U.S. executive director. Yet between 1977 and 1988 the board passed 113 loans on which the United States voted against or abstained, and U.S. directors do not necessarily follow congressional wishes (Brown 1992, 191–246; Stiglitz 2002, 51–52).

In the World Bank's official history, Catherine Gwin (1997, 195) suggests ambivalence in U.S. attitudes to the bank. "[W]hile it has supported the Bank for its capacity as a multilateral institution to leverage funds and influence borrowing countries," the United States has been "uneasy with both the autonomy on which the Bank's development role depends and the power sharing that accompanies burden sharing." Critics have argued that the World Bank Group and regional banks (the African Development Bank, the Asian Development Bank, and the Inter-American Bank) are not so much geared to their borrowers as to their donors' returns in the form of consultancy contracts, equipment sales, etc. (Payer 1982, xiii).

Figures are scarce, but between 1977 and 1982, U.S. firms received $3.4 billion in World Bank contracts, the majority going to consulting and engineering firms (*Washington Post*, February 6, 1984). From 1992–93, U.S.

companies gained a 14 percent share of World Bank contracts totaling $24 billion, compared to 8.5 percent for Germany and Japan and approximately 7.5 percent each for France and the United Kingdom (U.K. Department of Trade and Industry 1993). Gwin (1997, 273) calculates that until 1992 the United States received more than $42,000 billion in World Bank procurement against an expenditure of $9,564 billion in contributions to the IDA. The bank's expenditure of more than $1 billion of its administrative budget (excluding capital projects) annually in the Washington, D.C., area also brings it U.S. domestic support.

The Early Years of the World Bank's Privatization Thesis

The World Bank took up the ideology of privatization before the 1980s Western conservative governments came to power. Paul Mosley argues that the catalysts for the "new political economy" of program lending were internal organizational needs and Robert MacNamara's impending retirement (Mosley, Harrigan, and Toye 1991, 23–25 and 29). Developing-country debt reduced lending on infrastructural projects and threatened the bank's rationale (Lancaster 1993; Gwin 1993). MacNamara was also concerned about his legacy.

During the postwar decolonization period the prevailing concept of "development" centered on modernization and the lack of capital available to developing countries. Economic nationalism that often followed political independence included nationalization of foreign-owned enterprises or a range of controls over foreign investors often suspected of serving other than national interests. The World Bank's project-based finance demonstrated a Keynesian bias in loans to governments.

Then during the 1960s and 1970s the developing countries promoted their own economic and political ideology that substantially blamed the industrialized world for their economic plight. "Dependency" theory legitimated the newly established states as the most important actors in "development" (see, for instance, Frank 1966; Galtung 1971). Meanwhile in Asia the Japanese and other governments were beginning to use the state in a developmental capacity for export-led industrialization.

The introduction of the new World Bank instrument—the structural adjustment loan (SAL)—came against the backdrop of the second oil shock in 1979. There was already considerable organizational knowledge of such loans and conditionality that had been attempted in the 1960s and 1970s in India and Africa. Operations Vice President Ernest Stern, an American, and Senior Adviser on Structural Adjustment Lending, Stanley Please, had both been involved in unsuccessful attempts to impose economic liberalization on India (Mosley, Harrigan, and Toye 1991, 29). Stern and McNamara are said to have arranged the SAL program on a flight in late September 1979; the World Bank board gave reluctant approval in February 1980. The loans would provide finance over a period of several years in return for reforms in trade protection and price incentives. The World Bank made its first formal structural adjustment loan to Turkey in 1980.

The World Bank's Privatization Thesis—1980s

The World Bank's major concerns with the African public sector were the proportion of expenditure and employment it comprised, and its inefficiencies (World Bank 1981, 5). A shift in policy to privatization became publicly visible in the 1981 Berg Report on Sub-Saharan Africa. African heads of state objected that the report failed to address problems associated with privatization and gave "rise to the feeling (however faint) that ideological inclinations might have inadvertently slipped through" (cited in Browne and Cummings 1984, 165–66). But the Berg Report marked the beginnings of a new cross-conditionality between IMF, World Bank, and bilateral donors. It became impossible for developing countries to access loans without both Bank and IMF approval.

The two oil price hikes of the 1970s, the liberalization of financial markets in the 1980s, and unregulated offshore petrodollars led to overexpansion of lending by the U.S. commercial banks. U.S. interest rate hikes then led to a major debt crisis. SALs allowed the World Bank to use IDA money to repay IMF or commercial debts, to roll over loans, or to allow commercial banks to reduce their liabilities.

McNamara was succeeded in 1981 by Alden Winship (Tom) Clausen, previously head of the Bank of America—then the biggest commercial bank in the world. After five years of U.S. antagonism toward the World Bank, Clausen is said to have made it clear to the "Bank's top executives at the outset that he had no intention of maintaining his predecessor's focus on poverty alleviation." Clausen was reported by Mahbub al Haq, a McNamara adviser, to be adamant that "the only constituency that mattered was the United States" (cited in Caufield 1996, 144).

Clausen, himself, was an advocate of private-sector foreign investment and later of privatization. He is quoted as saying that the World Bank and the IFC now saw themselves as "partners with the private sector in the economic development of the Third World" (Rowen 1987). Under Clausen, Stern's department became the center of neoliberalism, to which Clausen is said to have abdicated, rather than delegated, power (Fleming 1986b; 1986c). The bank's chief economist also changed in 1982 from Hollis Chenery, a lifetime development economist, to Anne Krueger, a free market enthusiast specializing in trade. It was under Clausen's regime that the Multilateral Investment Guaranty Agency was established to guarantee private investment against risks in the developing world, and the IFC, under Sir William Ryrie, began a five-year program of expansion of its private investments.

The hostility of the Reagan administration to the World Bank during the Clausen era, occasioned by its loans to Nicaragua and Vietnam, and IDA lending to India and China, softened in 1985, with U.S. Treasury Secretary James Baker's debt plan to rescue U.S. commercial banks from their overextension in Latin America. Reporting at the time, commentators suggested, "Instead of it being an institution whose principle task is to finance development projects in the Third World, Mr. Baker sees it becoming a lever by which the U.S. can shift the developing world towards the free market policies which President Reagan espouses so vehemently" (McRae and Brummer 1985). Following Clausen's departure, Barber Conable, a former Republican in the U.S. Congress, acted to counter the charge from donors that the World Bank was inefficient and drifting. Under a corporate-style, expensive reorganization, Stern's role was downgraded and it is said that many ideologues left the bank (Mosley, Harrigan, Toye 1991, 23–25 and 29). But the organization remained secretive and bureaucratic.

Conable's appointment in July 1986 coincided with the height of the debt crisis. The World Bank was actually on the point of taking in more funds from developing countries than it disbursed. Its net transfers were close to zero in 1986, moved into deficit in 1987, and remained there until 1989–90 (Fleming 1986a; Brummer 1990). At the same time, in competition with the bank, the IMF's role on the African continent expanded during the 1980s to counter the deteriorating terms of trade.

A World Bank review of "divestiture" in 1987 found that, with the exception of Bangladesh, the majority of known closures of state enterprises were in Africa. Most of the sales were of small state enterprises that had previously been in private hands (Berg and Shirley 1987, 20–22). Don Babai, a World Bank insider, suggested "an unevenness of attention to privatization in World Bank lending programs," depending on the borrower's desperation (Babai 1988, 269). It promoted the outright sale of operators to foreign capital in the very poorest of countries, such as Bangladesh and Uganda, but not in the more prosperous parts of Asia.

The year 1986 marked the beginning of serious pressure on the recipients of bilateral U.S. aid to privatize state-owned assets. Secretary of State George Schultz publicly backed privatization and demanded that Agency for International Development (AID) missions achieve two privatizations each year. They had "to submit detailed privatization plans and strategies to AID Washington. . . . In those sectors where the government neither privatizes nor allows competition, the mission is now required to cease all activity" (cited in Young 1988, 10).

The U.S. AID set up a semi-independent body, the Centre for Privatization, an offspring of six U.S. consultancy firms. By January 1988 it had handled assignments in thirty countries on fifty-five projects. Jealous of their U.S. counterparts, British consultancies, such as the Adam Smith Institute, successfully argued that the British government should support the U.S. government in its attempts to influence the World Bank and should orient British bilateral aid toward privatization (Young 1988, 15). Bank-financed privatization benefited consultants.

Under Barber Conable the World Bank worked hand in hand with multinational business. In the telecommunications field it held a joint seminar in 1986 with the U.S. equipment suppliers; in 1987 it set up its Foreign Investment Advisory Service in the IFC; and in 1989 it developed a framework for promotion of the private sector through the IFC (Pfefferman 1992, 46; Sussman 1991, 45; Montagnon 1989). By the end of the 1980s, the neoliberal "Washington consensus" between the World Bank, IMF, U.S. Treasury, and Wall Street had placed privatization at the core of development policy.

But, following on from the U.K. experience, privatization was now represented not as in the interests of Western capital but as the "indigenisation of economic power" (Young 1988, 3). According to the World Bank (1989, 185) privatization was a means of "empowering the people" in sub-Saharan Africa. In Russia, Poland, and Czechoslovakia vouchers were issued to the general population for the purchase of shares in former state enterprises—to be subsequently bought up by foreign companies for almost nothing. Overall gains and losses involved in privatization were not discussed prior to the event. Instead, privatization was reduced to the realm of the technical.

By 1990 the bank was under attack from the Left and the Right for the failure of its policies. Over the previous ten years the population had become poorer in fifty-three countries. In 1991 the Bush administration attempted to alter the World Bank's charter so that 50 percent of its loans went directly to private companies. Other donor countries opposed the United States, but Conable is said to have agreed that the future emphasis of the entire World

Bank Group would be the private sector (cited in Rowen 1991). The bank fought back against its critics. According to Larry Summers, then its chief economist, the fault lay with governments. He perceived African rulers as "Kleptocracies" (1992, 6–9). A condition of World Bank aid to Africa, but not to China, became "good governance," thereby giving the bank complete discretion over institutional reform (World Bank 1989; Williams and Young 1994).

The collapse of the Soviet Union's economy in 1990 produced a study for the G7, conducted by the World Bank, the IMF, OECD, and the newly created EBRD, suggesting a crash program to shift Russia to a market economy (Hoffman 1990). The sudden opening of the previous Soviet colonies to foreign capital led to something of a Wild West atmosphere. In 1990 direct foreign investment in the region was nearly $700 million, double the 1989 total. Ryrie of the IFC suggested that the massive changes required in Russia to turn it from a command economy to a market economy required some humility (Ryrie 1991). Instead, the Soviet collapse led to celebrations at the triumph of U.S.-style capitalism. But Big Bang privatization in Russia under Boris Yeltsin proved a disaster. Russia's economy and those of its previous satellites spiraled downward.

The proportion of World Bank loans made conditional on specific privatization targets rose from 13 percent in 1986 to 59 percent in 1992. But critics were gathering. By 1990 the total debt of developing countries had reached 106 percent of GDP in sub-Saharan Africa and was unsustainable. Criticism came from the internal 1992 Wapenhans Report, from environmentalists, from the U.S. Congress, and from previous World Bank employees (Graham 1993a, 1993b; Irwin 1990; Oxfam 1993; Tran 1994).

The World Bank was accused of ignoring the impact of its structural adjustment lending on the poorest and of preferring authoritarian governments to democratic. For instance, in the four years following Marcos's 1974 declaration of martial law in the Philippines, the bank and IDA extended loans and credits totaling 4.5 times the value of all WorldBank/IDA assistance in the thirty previous years (Sussman 1991, 51). It was also accused of presenting Chile's dictator Augusto Pinochet as the "star" of its privatization program. In reality, Pinochet's policy of selling assets "free and clear" (i.e., without their debts) had created a drain on Chile's central bank, estimated at 2 percent of annual gross domestic product (Livingston 1992).

Conable left the World Bank in 1991 to be replaced by Lewis Preston, the former chairman of J. P. Morgan, the New York bank. Preston refocused the World Bank's remit on poverty and its transformation from a lending institution to one offering advice rather than loans (Prowse 1991). In its 1994 development report the bank began to promote itself as a "knowledge based institution" (World Bank 1994, 2). Also its publication of *The East Asian Miracle* in 1993 began to marry the Bank's neoliberal views on markets and development with the Japanese government's continued espousal of a strong role for the state in development.

Following the election of President Clinton and facing congressional cuts to the IDA budget, the bank began to revise its policy of secrecy. However, its involvement in the Narmada dam in India produced a blistering attack in 1994 from U.S. Treasury Secretary Lloyd Bentzen (Tran 1994). In addition, the Clinton administration called for more emphasis on the private sector and an expansion of the IFC's activities. Subsequently, the World Bank created a new vice presidency for finance and private-sector development, and African governments were once more castigated for not privatizing enough.

According to the IFC, between 1988 and 1995 developing country governments earned more than $132 billion from the sale of state-owned assets, and 3,800 entities changed from public into private hands. Foreign investors accounted for 43 percent ($54 billion) of the total, of which $45 billion was direct investment. The majority ($22 billion) went to Europe and Central Asia, with Latin America and sub-Saharan Africa both receiving half that sum (Bouton and Sumlinski 1997).

The imbalance of foreign direct investment reflected both the risk to investors and the desire of governments. In 1995 the World Bank announced a new guarantee program to encourage private banks to lend money to infrastructure projects in developing countries. In essence, the bank was "offering structural adjustment insurance" (Caufield 1996, 291).

After the disaster of the bank's fifty-year celebrations that were hijacked by critical NGOs, in 1995 President Clinton bypassed the U.S. Treasury and appointed James D. Wolfensohn to the bank's presidency. The bank faced a crisis. According to its own figures, "private capital flows to developing countries increased . . . from $40.9 billion in 1990 to $256 billion in 1997. . . . In the same period, multilateral and bilateral foreign aid declined in relative importance from 57 percent of all net flows to developing countries to only 15 percent" (cited in Rich 2002, 28). Middle-income governments, such as that newly elected in South Africa in 1994, were not prepared to accept the numerous conditions attached to World Bank loans. Yet without the profits on loans to these countries, there would be no money to run the institution. Wolfensohn therefore opened dialogue with the NGOs, often sided with the bank's critics, began to decentralize the bank's staff into recipient countries, instituted a policy of openness, and attempted to reorganize the bank and to bring in fresh blood.

In the meantime, the creation of the WTO in 1994 began to pin back the bank's activities in trade (Vines 1998). Finally, in 1997, the bank did a volte face. For the first time since the advent of structural adjustment lending, it acknowledged that there was a role for the state in "development"—that the state could accelerate economic development. For the first time the mainstream World Bank promoted institutional capability and state regulation of markets (World Bank 1997, 61). Yet as Mark Berger and Mark Beeson (1998) point out, the bank's conception of an effective state was "grounded in an elite-centred approach to political and economic change which implicitly, if not explicitly, endorses authoritarianism."

The Concept of Privatization and the British Model

The Bank's "new political economy" of structural adjustment with its emphasis on privatization came out of a renewed linkage in the 1960s and 1970s between economics and politics. Anthony Downs's theory of voting (1957) and Mancur Olson's theory of group action (1965) were both based on the concept of people as rational economic actors looking for the payoff in any given situation. Their ideas became influential after the first oil shock of 1973, when Keynesian macroeconomics seemed unable to solve the problems of stagflation—low economic growth and high inflation—in the industrialized economies. In Britain monetarist economics came to the fore, positing theories that the public sector had grown too large, thereby "crowding out" private investment through its financial demands.

Privatization of state-owned enterprises could solve this so-called crisis of the welfare state and replace declining tax returns. Originally under bureaucratic control, telecommunications in Britain was separated from postal services in 1969 and run as a "public corporation." This structure gave quasi-autonomy in day-to-day activity. The corporation had a target rate of return on capital, but the bureaucracy retained controls over investment (Hills 1986, 116–17). The subsequent privatization of BT in 1982 was important in that it provided a model path—from bureaucracy to private company—for the World Bank and ITU.

The British Thatcher government "socialized" privatization. It offered shares at a discounted price to the general public, guaranteeing a return within twelve months to individuals and bypassing trade unions' opposition. For those governments fearing opposition to privatization it demonstrated "micropolitics"—the art of changing the social order through changing individual economic incentives. In effect, mass bribery could bypass "the entrenched forces, the interest groups" (Pirie 1988, 58, 61). Once privatized, the economic interests of those who benefited from privatization—primarily shareholders—would keep the entity in the private sector. And it was for that reason—to prevent a prospective Labour government from renationalizing BT—that the Thatcher government created the first sector-specific regulator, Oftel (Carsberg 1985).

Headed by former academic Bryan Carsberg, the agency was semiautonomous from government. For the first three years of its existence it was close to BT, sharing Christmas parties and aiming to maintain BT's profitability. Even as it was created, a division of opinion arose between those seeing the regulator as imposing economic rules without discretion and those who considered that regulation was a political mechanism to deliver national policy (Moran 2003, 105). Economic rulemaking became predominant. Imposing the new mechanism of a price cap over tariffs, the regulator allowed the company to raise access charges to residential customers, to charge more for local calls and more to rural than urban users, and to decrease tariffs on long distance calls and on leased lines. These changes became known as "rebalancing" of tariffs.

BT's sole competitor, Mercury (part-owned by Cable & Wireless), concentrated on long distance communications and large user companies with their 60 percent of network traffic, even attempting to bypass the less wealthy cities of Britain. Two analogue cellular operators, unregulated on the basis that two operators constituted "competition," provided service with almost identical tariffs for business and the wealthy. By 1987 price hikes and a decline in quality of BT's service resulted in public perception of unfairness in this regulation in favor of business interests, and Oftel's legitimacy came under threat. It then began to use its discretionary power to favor societal interests; from 1992 the duopoly ended and the network became more competitive.

This, then, was the first model for corporatization, privatization, liberalization, and regulation of telecommunications. Privatization was the practical outcome of a political ideology with its own perspective on history, its own vision of a future society, and its own theory of human nature, representing the economic interests of investors and large business. Privatization (without regulation) was to be adopted by the World Bank and the International Finance Corporation in its dealings with developing countries, whereas corporatization followed by privatization (with regulation) was the model pressed by the telecommunications personnel of the World Bank on state-owned operators. Just as in the British case, competition and a concern for societal objectives were to come later.

Telecommunications Lending—The First Years

Although the World Bank began lending for telecommunications projects in the 1960s, their 2.5 percent share of overall bank lending reduced to 1.5 percent in the 1980s. During its first twenty-five years the bank made loans and credits for one hundred telecommunications projects of about $3 billion to forty-eight countries, providing maybe 25 percent to each project. Such low funding was partly explained by the sector's potential profitability, which excluded it from Bank financing. But also "many lending sectors at the World Bank" remained "unenthusiastic about telecommunications lending as a development priority" (Goldschmidt 1984, 180). In addition governments varied in their enthusiasm. By the late 1980s some countries, such as India, Ethiopia, Thailand, Nepal, and Kenya had had multiple loans, but most had had only one (Joshi 1987, 69).

The balance of finance for capital intensive projects requiring large amounts of foreign exchange came from internally generated funds (usually 45 percent), supplier and export credits (25 percent), bilateral aid (25 percent), and government sources. According to Robert Saunders, chief of the World Bank's Telecommunication Division, supplier credits could lead to corruption in procurement, expensive equipment, and dependence on expatriates (Saunders 1982, 486). The use of manufacturers' credits also led to many different types of equipment in the network (in 1990 Poland had fifteen types) with concomitant training and maintenance costs.

Particularly during the early 1980s, the bank advertised its expertise in competitive bidding for procurement and in the syndication of finance as leading to better financial terms (Joshi 1987, 7; Wellenius 1986, 4). The first book on telecommunications and development by World Bank staff was published in 1983, a forerunner to the ITU's 1985 Maitland Report. It attempted to convince developing countries that better telecommunications could contribute to economic growth (Saunders, Warford, and Wellenius 1983, 4–5).

The book did not extol the virtue of privatization. It concentrated rather on the issue of the autonomy of operators from day-to-day interference by government. Provided that the operator was efficient and had adequate resources, the authors actually preferred monopoly provision in small- and medium-sized developing countries. They specifically rejected the thesis that private ownership "would alone determine the operating efficiency or responsiveness of its management" (Saunders, Warford, and Wellenius 1983, 59–60 and 287).

But there were aspects of the neoliberal agenda that they adopted. From 1981 the mainstream World Bank had argued that users should be charged according to what they would pay for a monopoly utility—or "user pays," in the vernacular (Payer 1982, 108–10). Applied to the telecommunications sector the World Bank and U.S. AID wanted telecommunications operators to charge high up-front access charges to consumers so that tariffs covered all production costs and a surplus to cover expansion—i.e., cost-based tariffs (Saunders 1982). But although they accepted that the result would be large price hikes, there was no knowledge of how these could be politically managed or of their impact (Wellenius 1984a, V.3.2.4).

For instance in 1982, following a World Bank loan to the Philippines, the main operator's access charges doubled to US$415 for a residential user, when GNP was US$783 per capita. Bank lending was geared toward business rather than wider access. As Gerald Sussman observed (1991, 56–61) among the forty-five countries that received Bank assistance between 1962 and 1988, only five achieved teledensities of ten telephone lines per one hundred people or above.

Because of the bank's secrecy, rather little material is available on its first telecommunications interventions. However, based on leaked internal documents, Sussman argues that by 1980 the bank's interest in providing financial assistance to the Philippines was based explicitly on taking direct control over the country's communications board, a regulatory body established in 1973 by the dictator, President Marcos. The bank regarded it as too weak to provide "adequate standards of service" (World Bank/IFC 1980, 31, cited in Sussman 1991, 58). This is the first documented consideration by World Bank telecommunications personnel of a sector regulator.

Writing in the IMF's journal, Björn Wellenius (1984b, 36) began to argue for private sector involvement in the sector, in particular in the terminal equipment market. But he thought that "privatization of telecommunications infrastructures . . . offers no easy cure to the sector's ailments nor does it relieve governments of the need to make hard policy decisions and maintain a competent regulatory capability." At this time the bank's conditions for financing included the separation of the operator from government bureaucracy, rebalancing of tariffs, expansion of the network to rural areas, and other organizational improvements, but not outright privatization. Although the bank did so in other sectors, its telecommunications staff did not directly intervene to have senior management replaced (Wellenius 1993, 24n38).

After the intervention of the Reagan government, Richard Stern, Saunders's successor, stated (1986, 168) that before financing a project proposal, "We in the World Bank will often require our clients to take remedial action" to address any problem identified as a result of bank's analysis of organizational structure, tariff levels, and human resource development. In addition the bank would ask "with increasing frequency what is the government's policy with respect to monopoly or competition?" Among a list of pressures on developing countries to restructure their telecommunications, he acknowledged those from "the emergence of enterprises seeking telecommunications investment opportunities outside their own countries." For the benefit of these interests the bank insisted on freedom to repatriate profits and refused provisions imposing the use of domestic labor (1986, 170–71).

However, state owned operators did not welcome privatization. A report of a meeting with a group of PTOs in 1989 indicates some consensus on the need for regulation but not on privatization: "Seminar participants spent some time debating the definition of privatization and the continuum of organizational change from government department through a statutory corporation to a joint stock company (publicly or privately owned). . . . It was . . . noted that political, cultural and ideological considerations would primarily determine where such a particular country would find itself in this respect" (Stern 1989, 126). The authors saw the problem for developing countries as one of bringing market forces into the sector so as to balance efficiency with legitimate political and social considerations (Nulty 1989, 17–18).

Domestic resistance to PTO privatization during the 1980s was widespread. So, for instance, Malaysia first created Syrakit Telekom Malaysia as a public corporation in 1984 in order to develop the Malaysian ethnic majority (Bumapatra). But then, because of labor opposition, it took six years for the shares to debut on the Kuala Lumpur stock exchange. Other attempts at privatization, such as those in Thailand, Colombia, South Africa, Uruguay, Brazil, and Argentina failed (Petrazinni 1996, 42).

In Latin America as countries struggled with debt, the bank's policy was outright privatization of network operators. Chile was the first to react to its worsening economic situation. In 1988 General Pinochet sold 50.1 percent of the shares in the monopoly operator, Companiá de Teléfonos de Chile (CTC), to Alan Bond, an Australian entrepreneur. Pinochet defused opposition by selling shares to the public, then sold CTC debt-free. When Bond's company became debt-ridden in 1989, his shares went to Telefónica de España, the monopoly operator of Spain's domestic network, which also bought the largest shareholding (20 percent) in Entel, the Chilean monopoly satellite operator.

As Wellenius concedes, the World Bank's interests in Latin America co-incided with those of primarily U.S. commercial banks, seeking to exchange their poorly performing debt for performing assets (Wellenius and others 1993, 4). Debt/asset swaps formed the basis for privatization of the monopoly Argentinean operator, Entel. Again, success came under the Menem government with side payments to interest groups (Molano 1997, 102). In this and other privatizations the bank produced good retirement packages so as to defuse opposition (Petrazinni 1996, 47n33).

Argentina's Entel was divided into two regions and the auction produced only two bidders. Entel Sur went to a consortium led by Teléfonica and including Citybank, while the consortium buying Telco Norte included state-owned Stet (Italy), state-owned France Câble et Radio (a subsidiary of France Telecom) and J. P. Morgan. But then the consortia sold the shares on the open market for more than three times what they had paid. Two World Bank reports subsequently concluded that the major beneficiaries of privatization were foreign investors (Galal et al. 1992b, 9; Hill and Abdala 1993, 17–18, 31, and 39). The bank blamed the windfall profits on failure to have a regulatory body in place before privatization (Hill and Abdala 1993, 35–36).

In fact, in Argentina the bank had attempted to set up an independent regulator for the sector. But the regulator failed to act, threatening to bring all future privatizations into disrepute. A report of March 1993 concludes, "Autonomous regulatory agencies administering fundamental legislation are so alien to Argentina's recent history that it will probably take years to develop a "demand" for good regulation. For example, the autonomy granted the National Telecommunications Commission was originally abused, the members of the commission essentially did little work and its executive—at the suggestion of the World Bank—finally had to intervene" (World Bank 1993a, 22).

Despite this setback and the realization that a lending institution could not give the ongoing support required for institutional change, the bank's telecommunications staff continued to be "instrumental in assisting the development of regulatory frameworks" (World Bank 1993a, x and 17). It learned that "where privatization was completed very quickly, such as Argentina, Mexico, and Venezuela, the Bank was left without specific leverage on the much slower but critically important stages of developing regulation and competition" (Wellenius and others 1993, 24n40).

The World Bank's interests in promoting privatization tied in with those of a number of Western PTOs looking to invest overseas to offset threats to profits from potential liberalization of domestic markets. In the words of one commentator: "Privatisations are providing telephone companies in Europe, the U.S. and southeast Asia with an opportunity to off-load some of their monopoly profits . . . [including] the U.S. Baby Bells, Cable and Wireless and the Spanish operator, Telefonica" (Newman 1992). The IFC also financed France Câble et Radio in its African investments and the transfer of control of Cap Verde's operator back to Portugal Telecom in the early 1990s. In other words, the IFC reinstituted French and Portuguese control in their previous colonies.

In fact, consumers in industrialized countries were paying excess tariffs in order to finance foreign investments and shareholder gains. In 1995 a British financier commented that BT's dividends had grown markedly as a percentage of sales and on a per-line basis as had those of Tele Denmark, Telefonica, and STET (Heyworth 1995, 10). Henry Ergas, an OECD economist, commented, "Even in supposedly open markets, telcos still think they can cover losses by hitting consumers' wallets. And perhaps they are right. Blinded by the current rhetoric, regulators may continue to shelter them from the real competitive rough and tumble that would force high profit margins away" (Ergas 1994, 14). Regulators in Western industrialized countries were seemingly biased in favor of the dominant operators. Monopoly profits funded overseas adventure.

Although telecommunications personnel in the World Bank pressed for sector regulation, some in the IFC with a short-term focus applauded the greater potential for profits when countries failed to regulate (Ambrose, Hennemeyer, and Chapon 1990, 30). From 1990 and the fall of the Berlin wall, the IFC and other lending agencies became involved in Central and Eastern Europe, where, according to the PA Consulting Group, foreign operators' interest was mainly in how quickly they could achieve revenue growth (i.e., gain monopoly profits) (cited in Horten 1992). Reflecting its increased importance, from 1992 the IFC was reorganized and employed six specialists in telecommunications (Wellenius 1993, 25n47). Its goal was short-term private investment. In 1990 it favorably cited first generation cellular operations by U.S. West and Bell Atlantic in Czechoslovakia and Hungary where "no official approval is necessary for setting or raising cellular rates" (Ambrose 1990, 30). The result was that the U.S. companies pitched their prices so high that only Westerners could afford the U.S.-West/Bell Atlantic cellular tariffs in Poland (Horten 1992).

Despite the views of the World Bank's technical specialists, privatizations of PTOs went ahead without the establishment of regulatory agencies. For instance, Joseph Stiglitz says that in 1997 the sale to France Câble et Radio of the Côte d'Ivoire operator, including monopoly licenses for both landline and cellular, was done "as is often the case, *before* either an adequate regulatory or competitive framework was put into place. . . . The private firm raised prices so high that, for instance, university students reportedly could not afford Internet connections, essential to prevent the already huge gap in digital access between rich and poor from widening even further" (Stiglitz 2002, 56; emphasis in the original). Sometimes the push for privatization produced bizarre results. In Zambia the 1994 Telecommunications Act, setting up an independent regulator, copied word for word the British 1981 Telecommunications Act.

There was also a downside to regulation. Where governments responded to societal interests, potential buyers walked away. One example was that of Puerto Rico, where, in response to legislation that imposed regulation, buyers offered a low price for the network. Similarly, in "risky" countries commercial operators required not only high tariffs, but a guarantee that they could keep them at such a level for five or seven years—often politically untenable (Wellenius and others 1993, 23n24). In other countries in the early 1990s, such as Brazil, Columbia, and Uruguay, political opposition led to failure (Molano 1997, 3).

In response to these factors, attention turned to "cream skimming" the more profitable parts of networks. Entry into the cellular market or the building and operation of long distance lines involved better returns, less risk, and less political opposition than outright privatization of network operators. Hence, in Puerto Rico, following the failure of outright privatization, the government sold the long-distance network to Spanish Telefónica.

During the late 1980s and early 1990s the development of three regional standards—U.S., European, and Japanese—for second-generation digital cellular equipment promoted attempts by manufacturers to gain bridgeheads in Africa and Asia, so as to establish their standard as the "de facto" global standard. Manufacturers argued that digital cellular technology could replace fixed wireline technology for rural areas in developing countries. Such an argument, as advanced in South Africa, then legitimated entry by cellular operators into the urban areas of developing countries and the introduction of digital equipment that could not be manufactured locally. Manufacturers such as Motorola, with seventeen such investments in 1995, took equity in cellular operators in the hope of gaining equipment exports—a very similar strategy to that of ITT in the 1930s (ITU 1997b, 26).

During this period the World Bank's personnel argued that privatization had been a success. A 1992 study of privatization in Telmex of Mexico, CTC of Chile, and BT of Britain by the bank's Country Economics Department was used in subsequent World Bank publications to legitimize the policy (Galal et al. 1992a). In outline the report divided the politico/economic interests in each country into groups and then assessed the gainers and losers from privatization. By a judicious choice of dates from which to measure "welfare benefits," by failing to disaggregate domestic consumers from large business, and by ignoring costs of redundancies caused by privatization, in common with other World Bank papers, this report found privatization to be a resounding success (see Wellenius and others 1993, 8).

This report's overstatement of "success" was perhaps inevitable. As one insider explained, the Bank's policies needed to be seen as "right" in order to attract co-funding (Mistry 1989, 11). The 1992 internal Wapenhans report concluded that the bank's research agenda was largely directed "by the need to substantiate politically inspired shifts in policy direction" (cited in Caufield 1996, 302). The bank's publications were not subject to peer review and had to serve their employer (Caufield 1996, 302).

The early 1990s were problematic for the World Bank. The EBRD wanted Europe to itself. Middle-income developing countries rejected World Bank loans, and the Bank Group was subject to criticism from the ITU and developing countries for "yearly fluctuations by agency, by region, and by the ratio of telecom lending to total lending." The ITU claimed that "the amount lent by the International Financial Institutions for telecommunications totaled about 3% of all telecoms investment and less than the money recycled through the accounting rate system" (cited in Winsbury 1994, 28). For instance Kenya reported on a rural telecommunications project that cost $750 million, 90 percent financed by loans. Yet only 15 percent came from the International Financial Institutions; the majority came from export credits, commercial loans and supplier credits. The situation was as bad as the 1960s.

Yet in 1994 World Bank personnel warned developing countries that the situation would get worse. In total the bank estimated that $250 billion would be required in investment between 1995 and 1999, four times the level of investment in the 1970s and three times that of the 1980s. However, because only 10 percent of project finance would come from official sources, developing countries would have to generate 50 percent from private sources and 40 percent from internal cash generation. The bank argued that in order to generate the 50 percent required from private sources, developing countries would need to restructure and privatize their telecommunications sectors (cited in Winsbury 1994, 30). The following year the bank lent only $17.2 billion to telecommunications projects involving private participation (IFC 2004).

Because of the variation in indebtedness between countries and because of the structure of that indebtedness—whether owed primarily to commercial banks, multilateral institutions, or bilateral lenders—the World Bank's leverage varied from region to region. In Asia, where it faced competition from the Japanese-dominated Asian Development Bank, which emphasized the guiding role of the state in development, and from strong, authoritarian governments with plenty of funding opportunities, the bank was least influential. For instance, in 1992, Bangladesh rejected privatization of telecommunications as a condition of a loan from the Bank of Japan, which the World Bank would administer.

A major problem for the bank was that the countries of East Asia financed development from a high domestic savings rate and not from either loans or foreign investment. Indicative of this lack of success in Asian telecommunications markets in 1993, Björn Wellenius is quoted as saying, "No intention to reform, no money—that is likely to be World Bank policy in the future" (cited in Adonis 1994). However, in 1994, the terminology as well as the argument changed, specifically with relation to Asia.

Suddenly privatization was no longer the World Bank's preferred solution to lack of network expansion. Instead, referring to historical studies of the nineteenth century U.S. market, competition was the answer. But the competition advocated was not in long-distance provision (as in the United States, the United Kingdom, and others) but in local exchange service, where independent local telephone companies, such as those in the Midwest United States, would provide the solution to network penetration (Smith and Staple 1994, xiv and 26). This model, legitimized by American historical experience, was that imposed upon Hungary by the World Bank in 1993.

Following the disintegration of the Soviet Union, the Hungarian government became heavily reliant on loans. Privatization of MATAV, the Hungarian operator, promised a cash injection, and its investment advisers were in favor of outright privatization of a monopoly. The World Bank, with a different constituency to satisfy, decided that an alternative structure would increase penetration and opted for fifty-three local telephone companies. Subsequently, and predictably, French and U.S. independent telephone companies bought the most favorable licenses, leaving MATAV to serve the least profitable.

In 1994 this new policy in favor of U.S. independent operators and of equipment manufacturers was termed a "new pragmatism" (Smith and Staple 1994, ix). The idea was that a variety of private bidders would be attracted to provide rural service because of lower financial outlay. In fact, it took the financial crisis of 1997 for the World Bank to gain power in Asia outside its traditional clientele of the Philippines, Indonesia, and Thailand. The crisis shook the economies of Singapore, Hong Kong, and South Korea. South Korea was obliged to reduce its restriction on foreign investment in return for IMF/World Bank financing (Jeon 1998, 71). Indonesia, which in 1995 had opened up at the World Bank's behest to overseas operators on a build-and-operate monopoly basis in five regions, found those investors anxious to pull out (Indonesia 1998). Those countries, such as Malaysia, that came out relatively unscathed from the crisis eschewed the IMF and closed their borders to capital outflows. In other countries the cheapening of assets to foreign companies left a bitter taste, and a subsequent desire to build up reserves sufficient that they would not need the IMF or World Bank again.

Circumstances combined also to alter attitudes in Africa. During the 1980s, despite its leverage, the bank was unsuccessful in breaking through the strong African commitment to public ownership. However, in the 1990s competition for investment from Central Europe and Africa's lack of centrality in geopolitics promoted a fear among its governments that the continent might be falling off the international political agenda. When formal democratization in a number of countries lessened political uncertainty, Western European cellular suppliers fought for markets with the Americans and the Japanese in a rerun of the nineteenth-century carve-up of the continent—with its attendant corruption. "In some cases, new investors persuaded (often with bribes) governments to grant them special privileges, such as tariff protection. In many cases, the U.S., French, or governments of other advanced industrial countries weighed in—reinforcing the view within developing countries that it was perfectly appropriate for governments to meddle in and presumably receive payments from the private sector" (Stiglitz 2002, 97).

Following the election of President Clinton in 1994, Al Gore's Global Information Infrastructure initiative, and a new administration at the World Bank, there was a further shift in emphasis in telecommunications. Gwen Urey suggests that during the early 1990s "a campaign by telecommunications-related staff to get the board of directors to commit the Bank to formally articulating a telecommunications policy . . . led to the construction of some in-house history from the record of previous de facto policy" (Urey 1995, 119). Under the so-called "old agenda" in the pre-1986 years, it was said that the bank focused on the efficiency of the operator and funding of new investment. But under the "new agenda" the bank's telecommunications specialists were said to be concerned with "overall sector policies, regulatory environment, and institutional structures . . . [and with] advocating policies designed to promote new entry, competition, and private participation" (Communications & Development Team, 1994 cited in Urey 1995, 119). Whereas the World Bank previously collaborated mainly with the ITU, under the new agenda its partners were said to be the privately oriented IFC and the World Bank Group's Multilateral Investment Guarantee Agency (MIGA)—a politically acute rewriting of history.

In 1995 the new vice president of finance and private-sector development related a similar history at the inaugural meeting of the Global Information Commission, the private-sector side of the GII, in the creation of which the World Bank had been actively involved.

> First, at the bottom layer, we have for the past thirty years focused on getting the *basic telecommunications infrastructure* in place in our client countries . . . Then, as technological changes created the possibility for competition and private investment, we focused on the second layer: getting the *regulatory environment* right and encouraging reform, so that the private sector could come in and invest. This has been the main focus of our assistance over the past five or six years for more than 40 countries." (Rischard 1995, 51; emphasis in the original)

Regulation began to move center stage, but it was still handicapped by the overall view of the World Bank on the role of the state and the IFC's short-term outlook. For instance, the bank's 1994 Development Report (1994, 68) cites Malaysia's experience of telecommunications privatization to justify the conclusion that "moves toward privatization . . . need not wait for the formal creation of a comprehensive regulatory framework." Later research by the bank confirmed, however, that regulation before privatization held considerable long-term benefits (Wallsten 2002).

Perhaps not unrelated to the Clinton administration's GII focus, in order to celebrate its fiftieth anniversary, the Bank announced the creation of infoDev—the Information and Development Fund. The intention was to work with private companies, with bodies such as the ITU, and the institutions of civil society so that "the InfoDev Fund would be an operational vehicle . . . in the crucial areas of awareness raising, of advanced liberalization and regulatory work, and of promoting major applications of information technology, e.g. in the education, health and business services area" (Rischard 1995, 52). It was infoDev that funded the ITU's regulatory colloquium from 1997 and the provision of an online telecommunications regulation handbook. And with the signing of the Basic Telecommunications Agreement in 1997, the World Bank could no longer argue that the state was unimportant to development. Regulation, first promoted by the bank's telecommunications staff in the 1980s, had at last moved into the mainstream through the Regulatory Reference Paper of that 1997 agreement (see chapter 6).

The Impact of the World Bank's Privatization Thesis

From a low of $17 billion in 1995, World Bank lending for telecommunications rose to $39.9 billion in 1997 and $51.8 billion in 1998 as privatization increased in Central Europe (World Bank 2005). Overall the success of the bank was in promoting foreign direct investment into a wider spread of developing countries than were its traditional recipients. Yet patterns of foreign direct investment in telecommunications predominantly reflected previous patterns of colonialism, but with a twist toward the United States.

In Latin America, the U.S., French, Spanish, and British influence in the sector had a historical base. By the end of the 1990s, wherever network operations had been privatized or liberalized, they were run predominantly by U.S., Spanish, and French PTOs. Bell South ran cellular services in Argentina, Uruguay, Mexico, Venezuela, and Chile. In Argentina, in 1991 the PTO was split into two, the south run by Telefónica of Spain and the north by France Telecom and STET, the state-owned Italian operator. In Peru, from 1994 Telefónica owned the national operator Entel (Empresa Nacional de Telecomunicaciones del Perú). In Mexico, Southwestern Bell had a joint controlling interest with France Telecom in the dominant operator, Telmex. In Chile, from 1988 Telefónica controlled CTC (Compania de Telecomunicaciones de Chile SA) and STET controlled Entel (Empresa Nacional Telecom Chile SA—Entel). In Bolivia, from 1995 STET held 50 percent of the dominant operator, ENTEL. In Venezuela, from 1992 AT&T and GTE, the largest U.S. independent local exchange carrier, had a controlling interest in the dominant operator, CANTV (Compania Anonima Nacional Telefonos de Venezuela). A U.S. cellular company, Millicom, held a number of licenses in Mexico, Guatemala, Costa Rica, Chile, and the Dominican Republic. In Brazil, Motorola, Bell South, Bell Atlantic, and GTE were part of cellular consortia, while Telefónica, Portugal Telecom, Telecom Italia, MCI, and NTT of Japan were strategic investors in the two privatized Brazilian operators, Telebras and Embretel. But the Brazilian privatization in 1998, with its mix of regulation and competition, also brought in many domestic investors.

In Central Europe, European PTOs were slow to invest following German reunification and problems with the European exchange rate mechanism. US West Inc. had cellular licenses with Bell Atlantic in the Czech Republic, Slovakia, and Moscow. In total US West was involved in cellular service in ten cities in the former Soviet Union. Ameritech had a cellular license in Poland with France Telecom and in 1993 in alliance with Germany's Deutsche Telekom won thirty percent of the Hungarian operator Matev. The Hungarian GSM network licenses were awarded to one consortia led by US West, and another from the PTOs of the Netherlands, Denmark, Sweden, and Finland. A consortium of U.S. local telephone companies operated one of Hungary's local network licenses. Nynex had a yellow pages operation in the Czech Republic and also became the first Bell company to run a foreign system when it gained the license to run the Gibraltar network. Deutsche Telekom invested with Siemens in Moscow's GSM network and AT&T in Moscow's fixed-line services. Despite some European investment—for instance of Finnish Telecom in Latvia and Cable and Wireless in the Ukraine—the overwhelming impression was of U.S. dominance in Central and Eastern Europe.

Turning to Africa, the major private investments came from the previous colonial powers. In West Africa's previous French colonies, France Telecom's subsidiary, France Câble et Radio, held a proportion of equity in the international carriers of Togo, Madagascar, Chad, the Central African Republic, Djoubitie, and Equatorial Guinea, and cellular licenses in Chad, the Congo, Botswana, Cote d'Ivoire, and Niger. It was also the sole owner of the international operator in Cameroon. France Telecom invested in the fixed-line operator in the Central African Republic, Madagascar, Senegal, and Côte d'Ivoire, and in value-added services in Zambia. Cable & Wireless held a forty percent share in Sierra Leone's international carrier, ran the Botswana network, and in 1993 made a controversial investment in GSM in South Africa with M-Net, a subsidiary of US Viacom. Portugal Telecom held a cellular license in Botswana and ran the fixed network in Cap Verde. Swedish companies had cellular operator investments in Uganda, Namibia, and Ghana.

Although U.S. companies had previously had a presence in Africa in the equipment industry through ITT, particularly in Nigeria, cellular licenses gave them increased entry. Millicom obtained licenses in Ghana, Mauritius, and Tanzania, while Telecel Int. had licenses in Madagascar, Guinea, Burundi, the Central African Republic, the Democratic Republic of Congo, and Zambia. AT&T had a cellular license in Ghana and SBC held an 18 percent share in the privatized South Africa Telekom (ITU 1999a). Yet some African countries, including Guinea, Ghana, and South Africa, turned to a non-Western investor in Malaysia Telekom in the hope (to be disappointed) that the experiences of Western colonization would not be repeated. And in South Africa privatization and liberalization went hand in hand with the empowerment of local black elites.

The actions of the World Bank therefore opened up many developing countries to U.S. foreign investment for the first time. Although such opening must have pleased its major donor, the World Bank's major contribution came in its promulgation of nationally based regulation. Inadvertently, first through privatization based on national monopoly and state-based regulation and then through its policy of cellular competition, the bank contributed to a strengthening of the state-to-state system of international telecommunications, and a strengthening of developing countries' self-confidence that foreign direct investment could be controlled for national goals. Particularly in Africa, cellular licensees were often not allowed international gateways, protecting the national PTO. Inadvertently, the World Bank defeated the nonregulated end-to-end network aspirations of U.S. multinational users.

Conclusion

The World Bank was a U.S.-dominated international regulator of the domestic and international telecommunications markets of developing countries throughout the 1980s and 1990s. Because the bank was regarded as a U.S. institution, Congress and U.S. administrations were able to exercise direct and indirect power over the bank's personnel, over its conceptualization of "development" and its implementation. Surrounded by secrecy that enabled it to rebuff critics, it was U.S. bank officials who began to create an ideology of privatization in relation to developing countries predating either the Thatcher or Reagan Administrations. Its formalization into a specific policy in the 1980s can be seen as an institutional response both to threats to its own survival and to competition with the IMF. From 1985 privatization became part of the "Washington consensus"—a goal, not a mechanism—to be imposed on developing countries.

Until the end of the 1990s the mainstream World Bank coruscated the role of the state in development so that in Africa low-income developing countries were run either by the IMF directly or through NGOs. In Latin America the debt for equity swaps that allowed the takeover of Argentina's operator were later acknowledged to have not only allowed privatization at knock-down prices, but also to have benefited U.S. commercial banks. Privatization in Central and Eastern Europe similarly benefited foreign investors.

Without state regulation it was inevitable that public monopolies became private monopolies; that shareholders and large users benefited to excess. By 1987 feeble regulation in Britain had already made those beneficiaries evident and had led to a public backlash. Spread of that knowledge helped to promote debate and to delay privatization in Australia, Brazil, and Uruguay, among others. The bank's telecommunications personnel tried to establish a regulator in Argentina, but without success. While the bank demanded institutional reform, it did not have the time or personnel resources for the support required. Nor did regulation help in the sale of operators. In fact, the lack of regulation in the Central European states was an attraction to foreign investors. The weaker the state, the more likely it was to privatize outright because competition required greater administrative resources. Regulation traded off against the price that governments could realize—foreign investors at first wanted private monopolies and high returns in compensation for their assessment of risk.

The World Bank's own research and publicity material hymned the virtues of privatization. Yet in logical terms, the bank's policy was nonsense. Western, state-owned telecommunications operators or privatized monopolies, themselves inefficiently run, benefited from lax domestic regulation and milked their consumers at home in order to buy and supposedly run efficient telecommunications operators in developing countries. That, given the opportunity, they then ran those operators for the benefit of their domestic shareholders or government coffers should not have been a surprise.

Lending little of its overall budget to telecommunications, and even reducing that amount over time so as to pressure developing countries into restructuring, the bank's power was not all encompassing. Despite its best endeavors, many governments found domestic political opposition to privatization of the PTO too difficult and too costly. Had it not been that opening cellular markets to foreign investment was relatively costless for those governments, the bank's "success" rate would have been much less. One may argue, however, that the opening of cellular markets led to corruption and neomercantilism as well as to a rerun of nineteenth-century informal colonialism. To a large extent throughout this period it was the rural and urban poor who suffered while foreign investors and those local elites close to ruling parties benefited.

The World Bank itself was still handicapped in 1997 from the prevalent ideology of the 1980s. The bank's telecommunications personnel can be said to have flown under the radar of the mainstream bank in advocating regulation. The British model of a telecoms operator moving from bureaucracy to public corporation to privatization and liberalization was useful in plotting the path necessary for bringing telecommunications into commercial efficiency. An alternative model from the Midwest in the United States provoked much controversy when it was introduced.

Helped by the E.U. decision in 1987 to follow the U.K. experience and to separate out operation from regulation, the World Bank was helped also by its association with the ITU and the fact that from 1991 the two organizations pushed the same line. The ITU provided education to developing country personnel about regulation. The election of the ANC in South Africa also brought with it a determination to ensure that a privatized operator served national needs and an alternative locus of influence on that continent.

To a large extent, then—although the World Bank may have adopted the private investment mantra of successive U.S. governments and responded to U.S. indirect power by encouraging privatization through domestic monopoly—its impact was to strengthen the state-to-state system. By the time of the WTO 1997 Basic Telecommunications Agreement, it had spread knowledge of telecommunications regulation and the regulatory mechanisms that were available to control private investment in the societal interests of the country concerned.

6

GATT/WTO and
Telecommunications

This chapter looks at the way in which the United States introduced and pursued liberalization of the world telecommunications market in the General Agreement on Trade in Services (GATS), later the World Trade Organization (WTO). Originally the multinational user groups in alliance with the USTR were looking for the right to foreign investment and "Empire rules" networks without regulation from national governments. They gained these through NAFTA. But because of the FCC's traditional alliance with AT&T, WTO negotiations migrated to a "Western Union" model. Under different administrations the U.S. government used multilevel mechanisms—unilateral, bilateral, and domestic—to bolster its negotiating position and withdrew twice so as to satisfy domestic interests.

The chapter concludes that the final WTO agreement did not give the U.S. industry what it wanted. Because of its own protectionism, the United States weakened the Most Favored Nation clause of the GATT regime. And, in its demand for universal rules, the FCC-inspired Regulatory Reference Paper strengthened state-based regulation. Hence, GATS produced a strengthened state-to-state system with some liberalization. As a result, protectionism within the United States has increased. Together the FCC and USTR have fashioned bilateral agreements as spearheads for the next round of the WTO, and to limit the regulatory autonomy of other states.

GATT and the International Trade Organization

The General Agreement on Trade and Tariffs came out of the U.S. Reciprocal Trade Agreements Act of 1934, itself an amendment to the Smoot-Hawley

Tariff Act of 1930, under which Congress had increased U.S. tariffs on agriculture and manufactured goods. The Reciprocal Trade Agreements Act removed tariff setting from Congress to the Department of State and allowed bilateral negotiations to lower tariffs (Aaronson 1996, 21). But congressional protectionism meant that in 1940, 1943, and 1945 it renewed the act only by a small majority.

Then the Roosevelt administration thought of using the legislation to end British imperial preferences. However, the British and later the Canadians demanded multilateral negotiations and a reciprocal drop in U.S. tariffs. Because State Department officials had authority only for bilateral, sector-specific tariff reductions, the idea of multilateral reciprocity of tariff reduction was born. They opted for what they termed a "nuclear" solution, whereby a few trading nations would simultaneously negotiate bilateral trade agreements (Aaronson 1996, 48). This was the basis on which GATT was created in 1947. It was signed by twenty-three countries.

The State Department had originally planned an International Trade Organization through the U.N.-sponsored Havana Charter to establish a rule of law in the field of international trade. It envisaged an international body to enforce the national implementation. But the final charter was full of exemptions and, because 1948 public opinion had moved against trade liberalization, the ITO was never presented to Congress.

GATT, being nontreaty based, did not supersede national legislation, such as the Buy America Act of 1933, and could be presented to Congress as no threat to U.S. sovereignty. It became a vehicle for U.S. propaganda on free trade from the executive branch yet allowed the retention of protectionist legislation at congressional level. GATT was never about "free trade," only about reciprocity. Because of its origins in the "nuclear" option, in order to expand reciprocity from bilateralism into multilateralism GATT's central premise was its Most Favored Nation (MFN) provision, in which "any advantage, favour, privilege or immunity granted to any contracting party to any product originating in or destined for any other country shall be accorded immediately and unconditionally to the like product originating in or destined for the territories of all other contracting parties" (GATT, Article I [1]). In other words, any trade advantage given to one member state automatically applied to all. In order to end discrimination against foreign manufactured goods, part II article III also established the principle that goods imported from a member state had to be treated equally to those manufactured locally—the "national treatment" principle.

The agreement contained a number of "get-out" clauses for infant industries or balance of payments difficulties. Article VI also established the rights of states to take antidumping action by the imposition of countervailing du-ties. But the only overall exemption originally allowed was between those countries that had established a free-trade area. Then, in 1958, so as to limit Soviet influence, and because their markets were unimportant, members added Part IV (Article XXXVI [5 and 8]) to allow preferential treatment to developing-country exports (Watkins 1992, 33).

Political competition from the Soviet Union was also responsible for the inauguration of the United Nations Conference on Trade and Development (UNCTAD) in 1964 under the auspices of the U.N. General Assembly. The UNCTAD conference marked the creation of the Group of 77 developing countries and provided a forum for their pressure on GATT (Nye 1969, 336). Holding the major proportions of world trade, the United States and Britain were the most influential members in GATT's early years, but in the 1970s the European Union increased its influence (Curzon and Curzon 1969, 322).

The first five GATT negotiating rounds lowered the average tariff on manu-factured goods from over 40 percent in 1947 to approximately 20 percent in 1961. The agreement did not cover commodities. And from the 1960s, side agreements—the most famous of which was the 1974 Multi-Fiber Arrange-ment—protected U.S. and European manufacturing industries.

As tariffs on manufactured goods declined, nontariff barriers—quotas, standards, and procurement restrictions—became important barriers to international trade. By 1986 GATT had become an institution undermined by the West's bilateral and unilateral protectionist actions. To developing countries, comprising two thirds of GATT's ninety-two-country member-ship, it was a rich man's club. Hence the task of negotiators in 1986 was to restore GATT's legitimacy and to ensure that it better served its developing country members (Tyler and Dullforce 1986).

The Information Economy, Transnational Data Flows, and GATT

After the first oil crisis in 1973, Daniel Bell's book on the "post-industrial" society, positing a change from a manufacturing to a service-based econ-omy, was immensely influential, as was Simon Nora and Alain Minc's later French report on the information industry (Bell 1973; Nora and Minc 1980). Services became the new gold, and the 1974 Trade Act became the first U.S. trade legislation to include both goods and services. But the act was highly protectionist. It allowed the imposition of higher tariffs through Section 301 whenever an import reached its "peril point"—a certain proportion of domestic sales.

This, then, was the context in which U.S. representatives unsuccessfully pressed in the Tokyo round for the inclusion of services in GATT in 1975. In the following ten years, the U.S. commercial banking industry grew in importance as it recycled petrodollars into loans to developing countries. By the 1980s the liberalization of financial markets had made the twenty-four-hour global equity market a reality, and fast, secure, international communications essential. At the same time the penetration of data processing and the beginnings of competition in domestic communications markets increased the power of large users and data processing companies.

In the worldwide spread of Western companies, transnational data flows became linked to the re-emergence of the concept of the "free flow of information."[1] U.S. corporations viewed privacy legislation, passed by countries such as Sweden and Canada, as a nontariff barrier to trade (Eger 1981, 377). Developing countries also saw data flows as perpetuating economic dependency, but the majority had few resources to undertake the necessary regulation. Brazil and India were exceptions. Both developed an informatics policy. They linked state-sponsored computer industry development to UNESCO concerns about cultural imperialism and unbalanced information flows between North and South (Grieco 1984; United Nations 1983). U.S. multinationals regarded such policy as erecting barriers, not only to exports of U.S. equipment, but also to their freedom of telecommunications use (U.S. Congress, Senate 1983, 24).

In 1980, after lengthy discussions, the OECD issued voluntary guidelines on transborder data flows, primarily intended by the United States to prevent further protective national legislation. Peter Robinson, the Canadian chairman of the OECD Working Party, commented that many data users or providers had interpreted the term "free flow" to mean they "had total license to do whatever [they] wanted . . . regardless of the consequences" (Robinson 1985, 317). When it emerged in the 1940s, "free flow" was linked with the idea of end-to-end control of international telecommunications. Similarly, in 1981, the USTR listed access to flat-rate private leased lines, limits on private networks, and the use of standardization as nontariff barriers to "free flow"—all issues to reoccur in WATTC-88 and GATT negotiations (Schoonmaker 2002, 40). Financial companies were particularly concerned at the security risk implied in CEPT threats to force them off leased lines and onto public data networks (U.S. Congress, House of Representatives 1982, 132–35). However, the number of U.S. companies affected was small—estimates suggested that forty-two companies were responsible for 70 percent of the use of international leased circuits (O'Rorke 1985).

Although service exports were overstated, trade itself was of growing importance to the U.S. economy. But the coalition of farmers, manufacturers and organized labor that previously supported an open U.S. market was on "the verge of collapse" brought about by "recession, unemployment . . . and a sense of increasing barriers elsewhere" (Merritt and Cheesewright 1982). In the face of declining popular support and an increasingly protectionist Congress, the newly expanded USTR bureaucracy began looking for a domestic constituency to support free trade.[2]

There were a number of international institutions that the United States might have approached to take up liberalization in international trade. The United States had tried the OECD. But despite its $62 million annual budget and 2,000–person staff, the OECD's 1985 Declaration on Transborder Data Flows was weak, and the organization had no follow-up power.[3] UNCTAD, strongly influenced by the Soviet Union, was not an option. Yet, compared with the $183 million budget and 1,600–person staff of the IMF—put forward as a possible alternative—GATT had a small budget ($23 million in 1984) and only 100 professionals among 330 employees (Tyler 1984).

Schoonmaker (2002, 51) explains the U.S. choice in terms of GATT's linkage with free-market economics, but the reason may have been that GATT was an executive rather than advisory agency. In 1982, when the United States was institution shopping, Joan Spero, then of American Express, pointed out that GATT had a dispute resolution mechanism (Spero 1982, 152–55). The company was in favor of a "competitive, open international telecommunications marketplace," of domestic legislation "defining *rules* and creating *remedies*" against those countries that did not allow market access to U.S. companies. It also wanted to see bilateral agreements leading to multilateral rules and GATT as "a model for an information regime."[4]

In fact, GATT's dispute-resolution process was weak. Its tradition of consensus and its lack of a permanent list of arbitrators allowed delays and arguments. Its recommendations for action were also subject to challenge (Tyler 1984). Despite these weaknesses, the U.S. Business Round Table recommended the GATT Standards Code in 1985 as a possible model for a dispute-resolution mechanism in the field of international information flows. The ITU would remain the forum "for resolving technical issues" (Business Round Table 1985, 39). In contrast, academic Peter Cowhey argued that because the ITU and Intelsat regulated international telecommunications, GATT "should take away some of . . . [their] institutional competence." He suggested, "One of the keys to success for GATT negotiations is to pick up those issues that are more easy to remove from the ITU and Intelsat jurisdiction and keep those as a focus." The intention would be to mobilize "a coalition of interests of the industrial world" that would "have a strong interest in reforming the rules of governments in response to the information economy" (Cowhey 1987, I-4).

By 1987 companies such as the General Electric Information Services Company (GEISCO), Control Data, and IBM were lobbying for the introduction of services in GATT (Aronson and Cowhey 1988, 36–37). But it was the agenda set out by American Express, backed by the economic clout of internationally expanding financial institutions, that became the agenda of the U.S. government. The U.S. Coalition of Service Industries, led by Citibank, was reputed to be the driving force behind the U.S. efforts to open up the world market in services and liberalization of trade in telecommunications (Schniad 1992a). According to Carol Balassa, the first director of telecommunications services in the USTR, in 1987 financial institutions and enhanced network operators "were seeking the development of *legally binding and enforceable rules* to ensure that their entry into foreign markets would not be hampered by restrictions on the use of basic telecommunications services" (Broadman and Balassa 1993, 34; emphasis added). They wanted an "Empire rules" network controlled from the center.

Introducing Trade in Services into GATT

The 1982 GATT round came as power within the international economy swung away from developing countries. North-South discussions in 1981 in Cancun on a fairer distribution of global wealth ended with President Reagan's uncompromising call for liberalized markets and free trade (Dale and Chislett 1981). Yet, from the developing countries' perspective, the industrialized world prevented the export of their products through import controls and quotas while tying bilateral aid to their own exports (*Financial Times* editorial, October 26, 1981). Also, by endorsing monetarist policies, the GATT secretariat was perceived as Western-biased (Tyler 1984).

Criticized by Arthur Dunkel, the director general of GATT, for Congressional protectionism, the Reagan administration argued that to include services within GATT would allow the forging of a new domestic constituency in favor of free trade. But Brazil and India countered that trade in services was outside GATT's competence and should be discussed in UNCTAD as an investment, not trade, issue. They demanded that the industrialized countries honor existing obligations under GATT (Raghavan 1990, 72).

Such was the tide of protectionism from both the United States and the European Union that the GATT ministerial meeting of November 1982 became a test of its credibility. Before talks began the U.S. ambassadors to

GATT emphasized that the dilemmas of the developing countries should be taken into account (Smith 1982; Brock 1982). But the meeting developed into a series of bitter rows—between the United States and the European Union over agriculture, between the United Kingdom and the European Union over commonwealth preferences, and between north and south. Eventually, a communiqué was fashioned announcing a program of studies on services (Leeson and Jacobson 1983, 184).

The developing countries thought they had achieved a major victory because the United States appeared to give up on the introduction of investment issues into GATT. (In fact, the matter of investment rights was transferred secretly into the OECD.) The developing countries also thought that trade in counterfeit goods and intellectual property rights only applied to the Newly Industrializ ing Countries (Hong Kong, South Korea, Singapore, and Taiwan) and did not oppose their inclusion (Raghavan 1990, 72). From its side, the U.S. delegation was disappointed that it achieved only an agreement that individual countries would prepare papers on services trade (U.S. Congress, Senate 1983, 14).

The 1984 GATT ministerial meeting was characterized by U.S. threats to veto GATT's budget and Japanese refusal to allow U.S. officials to limit negotiations to OECD countries (*Mainichi Daily News,* August 23, 1984). In fact, with a growing proportion of world trade (26 percent in 1986), the markets of the developing world were of increasing importance to the industrialized. India and Brazil led developing countries in arguing that the industrialized nations should lift nontariff barriers that affected 30 percent of their manufactured exports and cease direct subsidies to agriculture (Dullforce 1985). But if developing countries did not open their service markets, they wanted to be sure that the industrialized countries could not take action against their manufactured goods. Brazil and India therefore pressed for services negotiations separate from those on goods. Again, much to U.S. disappointment, the outcome of the 1984 meeting allowed the secretariat only to prepare an "analytical summary" of a series of studies on services (Guest 1984).

U.S. Bilateral Strategy

Meanwhile the U.S. government was preparing an exit strategy from GATT—an option that then Chief of White House staff, James Baker, is said to have taken seriously (Odell and Eichengreen 1998, 191n15). The Reagan administration began negotiating regional free-trade agreements, the first in 1985 with Israel. Primarily to demonstrate political support, this agreement provided the beginnings of a framework for GATT and telecommunications.

In 1985 the United States also began negotiations with Canada. For Canada's Mulroney government, weakened by personnel scandals, by an imbalance of trade with the United States, and a deficit in manufactures, the agreement was of major importance. Canada had seventy staff servicing the negotiations but the United States only two (Niskanen 1987, 4). For the United States, the agreement was not a major priority but rather a lever against the European Union in the GATT.

As the Canadian negotiations began, U.S. Trade Representative Clayton Yeutter threatened to leave or to act outside GATT "if there was no agreement to start the new GATT round" (cited in *Financial Times,* November 15, 1985). The United States also threatened developing countries in the Pacific and Caribbean regions with cancellation of concessions under the General System of Preferences. Debt dependency dragged reluctant Latin American countries, such as Mexico, into the negotiations. And for the first time, in September 1985, the United States used Section 301 of the 1974 Trade Act to launch investigations into alleged "unfair trade practices" by Japan, the European Union, Brazil, and South Korea (Fleming 1985).

The USTR demanded that Brazil lift restrictions on U.S. companies in the micro- and mini-computer market and end the pirating of U.S. software packages. Sara Schoonmaker (2002, 86–87) traces the decision to the growing U.S. trade deficit in high technology and argues that "[t]he Brazilian informatics case was part of a broader struggle to establish the political terms for U.S. firms to enter foreign markets for computers, software, semiconductors, and other high-tech products and services . . . [and] extended . . . to include foreign investment and intellectual property." It is not clear whether the targeting of Brazil was directly linked to that country's procedural maneuvers in GATT, but the timing seems coincidental. Following the Brazilian government's decision to introduce a software law (designed both to placate the United States and also promote the domestic software industry), the USTR suspended parts of the investigation in December 1986.

The developing countries were soon split from each other by U.S. bilateral tactics (Raghavan 1990, 76). However, the GATT method of progressing via consensus, not votes, allowed Argentina, Brazil, Egypt, India, and Yugoslavia to continue to delay the introduction of services into the negotiations. At first, because France was unwilling to include agriculture in the round, they received backing from the European Union. But later the European Union offered liberalization of services as a tradeoff to protect French agriculture (Gowers 1985). It joined with Japan to influence the African Francophone bloc and the countries in the Far East. Japan also put pressure on India (McDowell 1997, 97).

In addition, "[t]he European Community spearheaded a campaign to inform a large number of African and Latin American countries that Brazil and India . . . were also reaping large surpluses on trade in services with other developing countries. It was implied that perhaps their opposition to change had something to do with the monopolistic rents they were earning in restricted markets" (Aronson and Cowhey 1988, 42). As a result of these pressures a group of twenty split away from the developing country bloc. Eventually, faced with U.S. threats that it would allow GATT to collapse if services were not included, the five major developing country opponents agreed to a preparatory committee "to determine the objectives, subject matter, modalities for and participation in the multi-lateral trade negotiations." The United States viewed this outcome as a "simple case of victory" (McDowell 1997, 100).

Bilateral pressures on developing countries ensured that the numbers opposing the United States gradually reduced, so that, by June 1986, when Chile and Colombia withdrew their backing, Brazil began to soften (Dawnay 1986). Once the preparatory committee began work, the Indian position shifted, also partly reflecting the ongoing domestic liberalization instituted by Rajiv Ghandi (McDowell 1997, 102). But a tactical alliance between the European Union and the Group of Ten, led by Brazil and South Korea, produced a compromise. Discussion on trade in services would take place outside the GATT Preparatory Committee, thereby preventing any bargaining between goods and services (Raghaven 1990, 77).

The Canadian Free Trade Agreement

The Canadian Free Trade Agreement was signed on January 2, 1988—the same day that "fast track" authority to President Reagan ended.[5] The agreement left untouched existing U.S. and Canadian trade laws. It allowed Canada to keep its monopoly of basic telecommunications service and a 20 percent limit on foreign investment in facilities-based carriers. It also allowed the Canadian authorities to oblige carriers to employ Canadian rather than U.S. facilities across the border. On the U.S. side it allowed the continuation of restrictions on investment in the radio sector and on public procurement under the Buy America Act. Although Canadian negotiators succeeded in exempting cultural industries, the agreement went further than that with Israel in providing a model for GATT (Mirus 1989).

Attached to the agreement was a Sectoral Annex on Telecommunications-Network-Based Enhanced Services. Although Canada used the Japanese type distinction between facilities-based and nonfacilities-based suppliers, the agreement used the U.S. regulatory division between "basic" and "enhanced." Its aim was "to maintain and support the further development of an open and competitive market for the provision of enhanced services and computer services within or into the territories of the Parties" (Janisch 1989, 100). It therefore opened up the Canadian market for enhanced services.

The agreement was unpopular in Canada. It was seen as "diminishing constraints on transborder data flows, restricting the scope of public sector activity, and limiting the range of telecommunications services that can be publicly regulated" (Winseck 1995). But for some members of the U.S. Coalition of Service Industries, the agreement did not go far enough. They wanted to gain access to and use of Canada's basic telecommunications services.

For the USTR these two free trade agreements provided the basis for a GATT negotiating agenda. Carol Balassa was to state: "The telecommunications agenda was subsequently adapted and expanded for the multilateral GATS negotiations in Geneva" (Broadman and Balassa 1993, 32). In particular, the USTR transposed to GATT the ideas on liberalization of investment and a telecommunications annex.

The Uruguay Round

In September 1986, the ministerial conference of seventy-four countries met at Punte del Este and agreed to launch the Uruguay Round. It excluded the Soviet Union whose formal application the United States opposed. Although the United States had succeeded in including four new areas—services, high technology, intellectual property, and foreign investment—developing countries had succeeded in the creation of two committees. The Group for Negotiations on Goods and the Group for Negotiations on Services both reported to a single Trade Negotiations Committee.

The GATS negotiations on telecommunications began at a time of almost panic at the decline of the U.S. domestic telecommunications market from a surplus of about $800 million in 1981 to a deficit of $1.9 billion in 1986. An upsurge in demand for customer-premises equipment had benefited Asian suppliers. Congress interpreted these changes to mean that the United States "provides foreign vendors with unrestrained access to the American market while their own markets remain closed" (Barr and Skelly 1987, 5). But this analysis was wrong. In fact, financial crisis and overvaluation of the dollar had hit developing countries where almost two thirds of U.S. equipment exports had gone (Neu and Schnöring 1989).

Nevertheless, following congressional pressure, in December 1986 the FCC issued a notice of inquiry regarding the steps it should take to bring foreign countries to open their telecommunications markets to U.S. firms under the "public interest standard" of the 1934 Communications Act. Commentators suggested that the docket was an obvious attempt to influence trade negotiations—a transparent "effort to 'jawbone' other countries to open their markets simply by the threat of FCC action" (Barr and Skelly 1987, 6). At the ITU's World Telecom Forum in 1987 the United States and the European Union each threatened retaliation against the other's protection of its telecommunications market (Feketekuty 1987, 189–93; Hardy 1987, 195–99).

GATT negotiations waited on the U.S. presidential election. Then, during the electoral run up, Congress created a populist trade bill based on bilateral reciprocity. Whereas Section 301 of the 1974 Trade Act dealt with specific goods, Clause 1377 of the 1988 Omnibus and Trade Competitiveness Act (known as Super 301) could be used to accuse countries of a broad range of unfair trade practices. Congress gave virtually no discretion to the administration. "Naming" of a country (based on complaints from U.S. companies) triggered a USTR investigation of up to eighteen months, and failure to come to an agreement within three years would spark retaliation. The USTR immediately initiated bilateral consultations with twelve countries on telecommunications equipment and VANs (Feketekuty 1988, 4). The provisions led to protests from the European Union that the United States was in clear violation of international trading rules and in its annual report for 1989 listed forty-two U.S. barriers to trade (Commission of the European Communities 1989).

The George H. W. Bush administration took office in January 1989 and appointed Carla Hills as the new U.S. trade representative. Following allegations by the U.S. Semiconductor Industry Association that the Japanese were "dumping" semiconductor chips on the U.S. market, trade talks with the Japanese were ongoing. In what was a secret agreement, the USTR pressured the Japanese government into cartelization of the world microchip market (Dryden 1992).

Employing techniques first used in these talks with Japan, the GATT negotiating team sought to broaden discussions to ministers of trade and finance within the OECD. A second "front" followed the bilateral line. The services agreements with Israel and Canada were to act as models for the GATT negotiations. A third "front" targeted the ITU and the aftermath of WATTC-88. A fourth "front" was to threaten trade sanctions on countries that refused to accept the U.S. negotiating line. Soon after her appointment, Carla Hills talked of using a "crowbar" to open markets and argued that "the credible threat of retaliation provides essential leverage in our market-opening efforts. Thus actual retaliation will be used, albeit reluctantly, to preserve the credibility of the threat" (cited in Riddell 1989).

True to this sentiment, the Bush administration provoked widespread indignation in May 1989 by naming Japan, India, and Brazil under Super 301, thereby directly linking it to the Uruguay negotiations (Dunne 1989a). Japan's problem was that Motorola wanted access to Nagoya and Tokyo for its U.S.-standard Code Division Multiple Access (CDMA) cellular technology (Tran 1989; Dunne 1989b). Both India and Brazil had continued to oppose GATT as the venue for negotiations on intellectual property. India also led opposition to U.S. proposals on rights of foreign investment for service providers. It proposed limiting the right of foreign establishment and requiring foreign companies to provide local employment and use local content. It also demanded transparency of information from the companies themselves (McDowell 1997, 114–15).

When the Uruguay round broke down over agriculture in December 1990, U.S. talk was of ditching GATT and concentrating on bilateral agreements. For their part Japanese negotiators were perturbed by the influence of U.S. industry advisers (cited in Thomson 1990). And for the developing countries there was frustration at their marginalization within the "Green Room" negotiations of the triad and the lack of transparency from the United States, the European Union, and Japan (*Third World Economics* 1990).[6]

The Telecommunications Annex

The United States, Japan, and the Republic of Korea put forward the idea of a telecommunications annex as a legally binding part of a services agreement in October 1989. One month later, Geza Feketekuty, the counselor to the U.S. trade representative, presented an unofficial document to a European conference. The draft annex envisaged progressive liberalization of national markets through a phased process of domestic regulatory reform to be reviewed annually by the GATT secretariat. Foreign providers would have the right to establish local facilities and the right to similar treatment to domestic operators. The annex would guarantee access to the network of the public monopoly on reasonable terms; would establish the right to leased lines and to interconnection of private networks to both private and public networks on a cost-justified basis; would allow foreign firms to use proprietary protocols and to transfer data across borders; would ensure monopolies did not cross-subsidize competitive services; and would allow foreign providers to participate in the standards-setting process. In other words, the annex contained a wish list of mechanisms that would prevent national regulatory authorities from exercising discretion over foreign investment. It represented the U.S. desire for an "Empire rules" system of end-to-end networks.

But the most important item was the indication that the United States would not accept MFN status for telecommunications. Because AT&T objected to the prospect of opening its domestic voice market, foreign operators would not have automatic rights of access for those services over which they had a monopoly in their own markets. Instead, each country would be allowed to establish special ground rules for the participation of foreign monopolies in their market. Each could negotiate preferential bilateral agreements with other countries that third countries could join (Feketekuty 1989, 6). The annex, therefore, envisaged first that domestic regulation within a country would be constrained by both detailed international rules and international oversight, and second that the GATS agreement would not apply to all telecommunications operators; in other words, the GATT principle of nondiscrimination (MFN) would be breached. Instead, there would be a series of bilateral agreements that could be progressively extended. This was a "Western Union" protectionist vision for the U.S. market coupled with "Empire rules" direct control.

The provisions in the U.S. document not only echoed but went further than 1989 proposals from the U.S. Council for International Business (cited in *Transnational Data And Communications Report* 1989, 5). In explaining the U.S. draft annex to the Group Negotiating Services, the U.S. representative stated, "The Annex was not intended to prevent countries from regulating telecommunications services so long as such regulation was consistent with the services framework."[7]

But the USTR proposals highlighted a conflict over the purposes of the proposed annex. Whereas India and others argued that its purpose should be to clarify the application of the GATS Framework to the telecommunications sector, the USTR wanted it to expand liberalization and impose centralized rules on domestic regulators. These views reflected alternative perspectives on market access and national treatment. The majority view was in favor of a "bottom up" approach under which each country negotiated "specific obligations . . . including conditions on entry and operation [to be] inscribed in individual country schedules." In contrast, the United States favored a "top down" approach where obligations to liberalize would be "of a general nature from which reservations would be made in specific cases."[8]

The U.S. draft annex drew resistance from both India, as leader of the developing countries, and the European Union, which contended that the annex was not needed. But for India it was a matter of both process and content. It objected to the closed-door side negotiations that had led to the draft. It also alleged that the U.S. draft "sought total liberalization of telecommunications as a mode of delivery without exchange of concessions." Then, when the Negotiating Group on Services had agreed to an informal telecommunications working group (chaired by Robert Tritt of Canada's Department of Communications) to consider whether a telecommunications annex was required, India hosted a meeting of eighteen developing countries and used the Technical Assistance Program in UNCTAD to take advice. Cameroon, Egypt, India, and Nigeria (later supported by Tanzania) produced two papers that made a distinction between basic telecommunications and its role as a "mode of transport for other economic activities." They argued that a commitment to allow access on the equivalent of "enhanced services" should "not entail automatic liberalization of basic telecom or other services" or demands for a country to upgrade or alter the regulation of basic telecommunications. They wanted developing countries to be able to place "reasonable conditions on access to and use of public telecommunications transport networks and services" and the "transfer of technology and technical cooperation" (Zutshi 1994, 24).

As the discussions were ongoing in 1990, the USTR named India under Super 301 for the second time. It complained that India forbade competition and "resale" in value-added services and discouraged the use of leased lines for data transmission (McDowell 1997, 144). Bilateral reciprocity and the GATT negotiations were firmly tied.

During the summer of 1990, the Telecommunications Working Party reached accommodation on the sectoral papers put forward by the European Union, Japan, South Korea, and India. The European Union text reflected the existing compromise around its services and open network provision directives between the liberalizers among member states (United Kingdom, West Germany, Netherlands, Denmark, and Ireland) and the more conservative, led by France. It made provision for universal coverage of all service sectors, full MFN treatment, and guarded against the abuse of power by dominant public and private sector providers. In contrast, the U.S. annex, with its demands for company rights to "resale" and end-to-end networks "was viewed by many as too extreme yet also too self-serving." It was around the European Union text that the chairman sought compromise (Woodrow and Sauvé 1994, 112).

By the time of the Brussels ministerial in December 1990, where the GATS agreement was to be signed, the draft telecommunications annex contained more square brackets than text.[9] In September 1990 the European Union had given notice that it wanted a "cultural exemption" for audiovisual services where "national cultural policy objectives" would override market access—a revisitation of the UNESCO debate of the 1980s.[10] Its draft annex on audiovisual services was supported by Australia, Canada, Egypt, Brazil, and Finland, but vociferously opposed by the United States and Japan. In turn, the developing countries objected to the clause in the draft telecommunications annex that imposed "cost-oriented" pricing of public telecommunications transport services. They argued that pricing was not mentioned in any other sectoral annex and was already addressed in the framework agreement (*Transnational Data and Communications Report* Jan/Feb 1994, 7). Eventually, following their continued opposition, the pricing clause appeared as a "best endeavor" clause in the draft telecommunications annex.

But then, just before the 1990 Brussels summit, under pressure from AT&T, the United States put forward a second Annex that excluded basic telecommunications from MFN. Telecommunications negotiations would be only about enhanced services. Although the draft telecommunications and audiovisual annexes were incorporated in the so-called Dunkel text presented to GATT Ministers, the texts were not adopted before negotiations broke down over agriculture (Dunkel 1992, 18). The first Gulf War then made negotiators unwilling to travel (*Third World Economics* February 1991, 5).

The Second Attempt at a Telecommunications Agreement

The Uruguay Round negotiations on a services agreement restarted March 8, 1991. By then, more than twelve countries and the European Union had submitted initial commitments. However, because of the U.S. withdrawal of "basic" services, with the exception of New Zealand, these commitments covered only enhanced services (*Transnational Data and Communications Report* 1991a, 5). The draft annex of 1990 was still the subject of disagreement. The U.S. delegation objected to clauses in the annex that contained "an inherent right to regulate."[11] It wanted to constrain or override national regulation with international rules and to gain better than national treatment for foreign operators—similar provisions to those in the secret OECD multilateral agreement on investment and NAFTA (see p. 191).

The U.S. delegation was facing pressure from U.S. business, which wanted an annex with similar provisions on leased lines, cost-based pricing, resale, and proprietary standards as that originally put forward by the United States—an "Empire rules" system (*Transnational Data and Communications Report* 1991a, 5–6) Citicorp objected to provisions in the draft Annex that had "all kinds of hooks for PTTs and regulators to limit access, usage and by-pass, in the name of safeguarding public service responsibility" (cited in Johnson 1991, 7). In particular two clauses stood out. The one stated: "Each Party shall ensure that no condition is imposed on access to and use of public telecommunications transport networks and services *other than as necessary*" (emphasis added). The other allowed that "a developing Member may, consistent with its level of development, place *reasonable conditions on access* to and use of public telecommunications transport networks and services necessary to strengthen its domestic telecommunications infrastructure and service capacity and to increase its participation in international trade in telecommunications services" (emphasis added). Both clauses allowed a discretionary "get-out" to developing countries.

But the major sticking point for the United States was the application of the central GATT principle of MFN to the telecommunications sector. Citicorp wanted basic telecommunications included, but AT&T wanted to preserve U.S. leverage so that MFN should not apply until state-based monopolies were restructured (cited in Johnson 1991, 6). In response, in 1992 the U.S. delegation presented a further modification to its own annex.[12]

The new text still exempted basic long-distance domestic and international telecommunications services from most favored nation coverage, but the U.S. negotiators now proposed that this exemption would not be necessary if countries with major telecommunications markets (particularly the European Union) made a number of commitments on liberalization. These were to include:

1. No limitations on the number of competitors permitted to participate in the basic long-distance services market
2. Permission for foreign entities to provide basic long-distance services through facilities-based competition, including the right to build, own, and/or operate domestic and international networking facilities, and through the resale of existing services networks
3. Permission for foreign investment in basic long-distance services
4. Affording to new providers, to ensure that they can operate economically, transparent, non-discriminatory, and cost-based access to services of basic telecommunications
5. Establishment by parties of a system of fair and transparent regulatory procedures administered by an institution with independent regulatory oversight (Reproduced in *Transnational Data and Communications Report* 1992, 5)

The United States demanded that all GATS signatories with a "major telecommunications market" should bind themselves from January 1, 1996, to full market access and national treatment in basic long-distance services (both domestic and international). Then the U.S. derogation would end because "other major telecommunications markets [would be] equally as competitive as the United States."[13]

In fact, several conditions in the U.S. annex were not actually fulfilled in the U.S. market itself. The U.S. document was aimed at the European Union, which had not then agreed to liberalize its voice transmission markets. The European Union replied in April 1992 that it would "give an up-front, unconditional commitment to MFN" followed by sectoral derogations, where necessary, but "the Community cannot accept that the U.S. unilaterally determines what constitutes a barrier or when 'mutually advantageous market opportunities' in telecommunications have been obtained. Nor can the Community accept U.S. efforts to negotiate under threat of unilateral retaliation. . . . In addition, such sectoral reciprocity is inconsistent with the principles of the multilateral trading system" (Commission of the European Communities 1992, 7). The United States would not agree to the Draft Final Act because it opposed MFN status for all signatories (Meyers 1992, 12). The negotiations came to a halt and GATT hit crisis.

NAFTA

Meanwhile the final draft of the NAFTA agreement between the United States, Canada, and Mexico was released on September 6, 1992. Because the United States was not prepared to allow an MFN clause within GATS, it also excluded MFN from NAFTA (Tiger 1992, 18). The agreement also excluded basic telephony and broadcasting and cable television. But it included much of the detailed wording of the first U.S. GATT telecommunications annex. This is not surprising since, according to a Canadian negotiator, "We simply transferred the conceptualization and, in many instances, the language from the GATS articles to NAFTA" (cited in Tiger 1992, 17).

NAFTA covered access to and use of public telecommunications transport networks or services, the interconnection of private networks, enhanced or value-added services, and standards-related measures. It contained a provision ensuring that private leased circuits were charged at a flat rate, not according to "cost." Canadian and U.S. firms were allowed to operate "intracorporate" communications both in and between the signatories but, unlike the GATS Annex, "intracorporate" in NAFTA covered all communications.

It also liberalized the movement of information, including access to information stored on databases in the territory of any party. Each of these provisions favored large users and resellers. NAFTA did, however, allow cross-subsidization between public telecommunications transport services, thereby allowing low-cost local calls to residential and rural users (Tiger 1992, 17).

NAFTA was also specific in the limitations it placed on national regulators—again, as the USTR had wanted in GATS. Information required for any license was restricted to establishing the applicant's financial status and technical conformity [Article 1303]. A regulator could not require a VAN applicant to provide its services to the public generally, cost-justify its rates, demand that it file a tariff, interconnect its networks with any particular customer or network, or conform to any particular standard or technical regulation. Under Article 1304 the regulator could only impose standards to prevent technical damage to the network, prevent interference, prevent the malfunctioning of billing equipment, and ensure users' safety. After one year, each party was to accept mutual recognition of equipment. A whole range of regulatory mechanisms was also specified to ensure that a dominant operator did not compete unfairly with a VANS operator.[14] In total, the agreement imposed centralized rules on regulators and favored the company over state sovereignty.

NAFTA was the first free trade agreement that placed private corporations above national governments. The corporation could sue the government, but not vice versa. The trade unions contended that the Canadian regulator had given up its right to regulate resellers or enhanced service providers or private leased lines, other than on grounds of technical threat to the network. They argued that, in effect, "transnational corporations will be able to provide their own data and enhanced service needs and to offer public telecommunications services." And provisions that prevented regulators from requiring resellers to base their prices on "costs" yet demanded that the operators of public networks had to price to reflect "costs" could result in foreign multinationals operating "a ruthless, price-cutting, money losing strategy" with similar results to those of the deregulated airline industry. In addition, U.S. companies based south of the border were to be allowed access to Canadian databases, posing a direct threat to Canadian jobs (Schniad 1992b). The United States had successfully imposed an "Empire rules" end-to-end system based on its own market divisions.

On the U.S. side of the border, opposition to NAFTA and the candidacy of Ross Perot were factors in the 1992 election victory of President Clinton.

He immediately instituted side-agreements to NAFTA on the environment and labor so as to placate environmental groups and labor unions. Mickey Kantor, President Clinton's former campaign manager, replaced Carla Hills at the USTR. But NAFTA had drawn attention to the potential impact of free-trade agreements. The Citizens Trade Campaign, representing a coalition of "consumer, environmental, labor, family farm, religious and other civic groups," mounted a campaign to educate the public "that citizen interests are central to U.S. trade policy" (Aaronson 1996, 140–41). NAFTA only narrowly passed Congress.

APEC: An Alternative to GATT?

Just as the Reagan administration had looked to the Canadian Free Trade Agreement to counter protectionist sentiments at home and bring Europe into Uruguay Round negotiations, so the Clinton Administration looked to bilateral and regional agreements. One possibility was to expand NAFTA to Latin America through the Free Trade Area of the Americas. In addition, the Asia Pacific Economic Co-operation Forum, created in 1989 as a result of a Japanese/Australian initiative, provided another potential regional grouping. Formed out of fear that the end of the cold war would see the United States pull out of the region, that a "fortress Europe" would isolate it economically, and that China, with its rising economic power, was not engaged in the region, the forum originally brought together twelve member economies including the United States and the members of ASEAN (Association of Southeast Asian Nations).[15] In 1991 they were joined by Hong Kong, Taiwan, and mainland China, then looking to join the WTO. U.S. involvement was sporadic until 1993, when for the first time it played host at Seattle.

In preparation for Seattle an economic plan by an "eminent persons group" under the leadership of the United States proposed a formal regional structure, rejected by the rest as attempted U.S. hegemony (Hellman 1997, 95). Instead, in 1994 APEC set itself to support the WTO process and set a target of free and open trade in the industrialized countries of the Asia-Pacific no later than 2010 and in the developing countries of the region no later than 2020 (APEC 1994). But, unlike NAFTA, which only offered concessions to those countries that had opened their markets to a similar level, APEC offered "open regionalism" to all comers. From the E.U. perspective, the U.S. engagement with APEC in 1994 demonstrated that it might walk away from the GATT negotiations.

The Third Attempt at a Telecommunications Agreement

The GATS negotiations stalled throughout 1992 because of the Presidential election. The European Union was also said to be too preoccupied with the implementation of the February 1992 Maastricht Treaty that presaged the inauguration of the European Union in November 1993. With the election of President Clinton it became urgent to conclude the Uruguay Round before the end of "fast track" authority in March 1993. That deadline was missed but negotiations on telecommunications began again.

According to Ben Petrazzini (1996, 12), "Well-documented folklore" has it that "basic telecommunications became a subject of negotiations because E.U. Commissioner Sir Leon Brittan 'challenged' U.S. Trade Representative Carla Hills to take up the issue. She agreed, even though AT&T, the regional Bell operating companies (RBOCs), and other players in the U.S. industry were not then ready for serious negotiations." In June 1992 the Swedish delegation, supported by the European Union, the United States, and others, proposed further negotiations on basic telephony. A draft protocol of November 11, 1992, proposed the establishment of a Negotiating Group on Basic Telecommunications (NGBT) with a two- or three-year mandate "to prepare a report on how basic telecom could be covered by GATS" (*Transnational Data and Communications Report* 1993a, 5). Once the existing GATS agreement was signed, the negotiating group would come into being. In addition to working toward a basic telecommunications agreement, the intention was to freeze bilateral negotiations so that countries could not undermine the multilateral agreement—a standstill constraint on the United States.

The language of the draft protocol caused dissension among U.S. industry actors with AT&T in favor and the U.S. Council for International Business against the proposed standstill. "Only after bitter fighting between AT&T and large users (with the RBOCs largely on the sidelines) could the negotiations go forward" (Petrazzini 1996, 12). The USTR was also against the standstill. Mickey Kantor told a Congressional subcommittee in August 1993: "We pursue market-opening opportunities in a variety of settings, simultaneously in bilateral, regional and multilateral negotiations." The USTR was promoting liberalization in the Uruguay Round, in NAFTA, and in bilateral trade relations with Japan (cited in *Transnational Data and Communications Report* 1993b, 5). Part of the strategy was to use agreement at one level to leverage agreement at another.

Re-entering negotiations on the telecommunications annex, U.S. negotiators pressed for inclusion of the NAFTA clauses giving unrestricted transfer of business-related information and a loose definition of "intracorporate" communications. However, time constraints prevented a wholesale U.S. renegotiation of the Dunkel text (Raghavan 1993a; 1993b). In particular, the United States and European Union agreed to disagree on the audiovisual sector. The European Commission, pushed by the French, took an MFN exemption on some audiovisual services.

The final annex began with a narrow definition of "intracorporate" communications that specifically excluded provision to third parties and to companies that were not related subsidiaries (similar to ITU recommendations prior to WATTC-88). It also excluded the NAFTA clauses on the pricing of public network tariffs and on flat-rate leased lines and defined value-added services narrowly to exclude "switching, signalling and processing functions." It also allowed signatories to specify protocols to service suppliers when "necessary to ensure the availability of telecommunications transport networks and services to the public generally." All these provisions were narrower than NAFTA. Only Article 4 on the transparency of regulation was almost the same as Article 1306 of NAFTA.

Five of the clauses specifically mentioned developing countries, and the final telecommunications annex of 1994 retained "get-out" clauses similar to those previously rejected by the United States. In addition, during the last stages of the negotiations, at the insistence of developing countries, the contentious clause on "cost-oriented" pricing was dropped from the annex (Zutshi 1994, 24). In a further clause, members agreed to promote international standards through the ITU and the International Standards Organization and to consult with the ITU on matters arising from the implementation of the annex. Hence, rather than replacing the ITU, the final annex recognized the ITU's role as a standards and advisory institution.

Altogether, fifty-seven countries agreed to liberalize their value-added telecommunications markets to varying degrees at varying times; the GATS provided a "framework" document laying out principles only applicable to the schedule of commitments made by each country. Few signatories were prepared to liberalize further than the status quo.

The New Institution of the WTO

The GATS agreement setting up the World Trade Organization was signed in Marrakech on April 15, 1994. Detailed proposals for the new organization originated with the European Union, Canada, and Mexico in late 1991. The first public intimation of the organization came in a leak to Public Citizen, the

U.S. consumer group established by Ralph Nader. Subsequently, 160 NGOs dealing with development protested to sixty governments against the draft WTO. They argued that the new body was to have a legal personality placing it on a par with U.N. institutions but not under the United Nations, and would have "no commitment to sustainable development or environmental protection or allow an adequate process of democratic consultation." In addition, they argued that the intention to allow the Multilateral Trade Organization, as it was then termed, to act as a trade policeman in conjunction with the IMF and World Bank was unacceptable (Lang and Hines 1993, 48–49).

Fearing congressional backlash, despite the presence of the (then) U.S. ambassador to GATT in the informal group that agreed with the Dunkel Text of December 1991, the United States proposed a so-called GATT II that would have allowed "grandfathering" of domestic law (i.e., allowed existing measures of protectionism and bilateral reciprocity). The European Union refused (Raghavan 1993c). The U.S. delegation eventually agreed to the WTO only on the last day possible, and then only when the name was changed from "Multilateral Trade Organization." As a result, there was no U.S. domestic debate (Odell and Eichengreen 1998, 203).

The World Trade Organization, subsuming the GATT staff, would be governed by a ministerial conference composed of all member countries to meet at least once every two years, a general council and three subordinate councils—one for goods, one for services, and one for trade-related aspects of intellectual property rights. Each member country would have one vote, but the tradition of consensus decision-making established under GATT would continue. A ministerial decision at Marrakech also reaffirmed the WTO's relationship with the IMF on the same basis as that previously governing GATT.[16]

At the time of the WTO's inauguration, Director General Peter Sutherland publicly expressed his ambition to see the WTO take the role assigned to the International Trade Organization in the 1940s (Williams 1994). But the United States' suspicion of an international organization that might compromise national sovereignty led to a small organization. In 1996 the WTO had only 513 staff members and a budget of $94 million, compared to the World Bank's 7,000 staff members and budget of $1.375 million. Richard Blackhurst contends that the major industrialized countries deliberately kept the WTO's budget and staffing low so as to prevent it from gaining a leadership position in the field of trade, from providing technical assistance to developing countries, and to boost the influence of their own permanent delegations (Blackhurst 1998, 56). With such a small staff, the WTO also had an agenda overload, particularly in coordinating with other international institutions (Krueger 1998, 404n3).

One of the intentions of those promoting the WTO into a coequal Bretton Woods body may have been "to delimit what they claim to be encroachment of the IMF and World Bank onto the trade terrain" (Pipe 1993, 28). Promoting the "long-range balanced growth of international trade" (Article 1) was one of the main purposes of the World Bank, but the WTO could now claim the premier position on trade. Writing from a bank perspective, Julio Nogués suggests that the 1996 request from the WTO for formal collaboration with the World Bank and the IMF may have been an attempt to leverage increased resources (1998, 92).

Bank personnel had to be reassured that their "independence" of research would not be compromised by formal links between the two institutions. And the presence of WTO personnel at World Bank meetings with client countries was restricted. Nogués comments, "The multilateral character of the WTO will be significantly strengthened the day its members provide the secretariat with the resources to undertake critical and independent research" (1998, 89). David Vines (1998, 69 and 73) also suggests that the WTO was "a very peculiar organization" that did not have the capabilities of the World Bank in terms of policy leadership and research on trade liberalization. Developing countries also perceived the WTO to be biased against them, and there were calls for "an urgent review of the role of the WTO Secretariat and its relationship with both UNCTAD and the World Bank" (Hirsch 1998, 396). In other words, bank supporters and developing countries alike saw the WTO as subordinate to U.S. interests.

The Negotiating Group on Basic Communications

Negotiations on basic telecommunications began in an informal technical working group that met from September 1993. Participants included Sweden, the United States, the European Union, Japan, Korea, Canada, and Australia, but no developing countries. Discussion centered on what form the outcome would take—a protocol signed by ministers or a standard format for schedules of commitments on basic telecommunications services. The United States and European Union raised again the question of a GATS telecommunications committee, and again the United States put "accounting rates" on the agenda.[17]

The Marrakech ministerial of 1994 then established the Negotiating Group on Basic Telecommunications, open to all governments to negotiate "the progressive liberalization of trade in telecommunications transport networks and services." The group was to present its report no later than April 30, 1996, and there was a standstill on bilateral agreements until that date.[18] Neil McMillan, a British civil servant, became the chair.

The negotiating group agreed to consider regulatory reform, interconnection, structural and accounting separation, number portability, pricing policy, and "accounting rate" reform.[19] But illustrating the difficulty of these negotiations, the group had first to accept a common definition of "basic telecommunications"—a U.S. term nowhere defined at an international level. In total, the group met seventeen times. It began with twenty full participants (twelve European Union members counting as one member) and seventeen observers.[20] Numbers increased gradually to April 1996 when there were thirty-seven full participants and twenty-four observers. Early developing country participants included Chile, Cuba, Cyprus, and Egypt. Brazil and India both took observer status.

What became known as the "Regulatory Reference Paper" was an FCC-initiated project. According to Laura Sherman (1998, 18n53), in December 1994 U.S. negotiators convened a meeting on regulatory objectives. This group, known as the "Room A Group," chaired by the chief Japanese delegate, originally consisted of the United States, Australia, New Zealand, Korea, and the European Union. It put forward a paper to the negotiating group in February 1995 addressing interconnection of competing suppliers to the public network, "competitive safeguards" against dominant service suppliers, transparency of regulatory processes, and the independence of regulators. The U.S. representative "described . . . the paper as reflecting the U.S. experience in moving its domestic regime from monopoly to competitive supply."[21] But other negotiating group members expressed concern at the implication "that there was only one way to achieve deregulation and competitive entry" and that licensing and universal service requirements were not mentioned.[22] The European Union at first questioned the rationale for the paper but then accepted that the GATS framework agreement and the telecommunication annex did not cover the separation of regulation from operation, some aspects of licensing and interconnection, and the use of the term "competitive safeguards." Without naming the United States, the European Union objected to the use of this term to describe "measures which condition the market access of a service supplier of another Member to an examination of whether that other Member affords equivalent market opportunities in its territory." It was referring to the FCC's bilateral Effective Competitive Opportunity (ECO) test.

The ECO test was a precondition that foreign companies had to meet before the FCC would grant a waiver in the "public interest" of Section 310 of the 1934 Communications Act and allow them entry to the U.S. market. The inauguration of the test under the Clinton administration signaled the culmination of a USTR campaign dating from 1992 to bring the FCC under its aegis.

During the Bush administration the USTR had been successful in preventing the FCC's liberalizing its regulation of U.S. common carriers that were more than fifteen percent owned by a foreign telecommunications entity (Broadman and Balassa 1993, 36). The final FCC rule on what was termed "dominant carrier" status, issued in November 1992, agreed to a three-part classification. The effect was to give the FCC virtually unlimited power to block access by carriers in the interests of U.S. companies. FCC Chair Reed Hundt in 1994 used its provisions at the behest of AT&T to block British carriers becoming international resellers. It cited Oftel's failure to regulate interconnection and number portability (*Communications Week International* September 12, 1994, 4). Then, following the administration's success in pushing the WTO implementing legislation through a "lame-duck" Congress in December 1994, the FCC went a step further with the formulation of the ECO test.

This ECO test was the public declaration of U.S. intention to demand sectoral reciprocity within the ongoing negotiations. As Peter Cowhey, by then an FCC counselor (1998, 900n5) observes: "The Commission intentionally cast ECO in a manner designed to signal the rules it would consider necessary for a satisfactory WTO agreement." Yet, as the European Union and also the U.S. Council of Services pointed out, the concept of reciprocity was not compatible with WTO principles. Subsequent discussion in the negotiating group concluded that the "public interest" could not be defined sufficiently to be included in a GATS offer as a reason for an MFN exception. In fact, the ECO test was to cause ongoing problems within the negotiations.

In its reaction to the U.S.-based reference paper, the European Union also pointed out that tariffs and costs (including accounting rates) had been excluded from the telecommunication annex after fierce opposition from developing countries. But there were not enough developing countries in the negotiating group to prevent the issue returning to the agenda.[23]

As discussion continued on regulatory principles, the United States—supported by Canada but opposed by the European Union—put forward its view that the national regulator should not only be separate from the operator, but it should also be separate from the government office that had an interest in the public network operator. This was the first expression of later U.S. demands that governments should not own network operators. Because there was no general agreement, the discussion moved once again to informal meetings. Later sessions of this "Room A Group" were attended by representatives of Brazil, Singapore, Chile, Mexico, and the Philippines (Sherman 1998, 30n54).

The formal offer made by the United States in July 1995 contained within it a set of regulatory principles that it committed itself to adhere to if the negotiations succeeded. Its offer, however, applied only to interstate and international basic services (i.e., federal responsibilities) and retained domestic restrictions on the commercial presence of foreign companies in terms of radio licenses, access to Intelsat and Inmarsat, and rights to land submarine cables. Its offer was also made contingent on a "critical mass" of WTO members committing to market access and national treatment and to procompetitive regulatory disciplines.[24] When asked how he defined "critical mass," the FCC's Scott Blake-Harris answered without irony: "What we want to do is eliminate all governmental barriers to foreign investment and foreign competition. It is in the best interests of consumers and businesses all over the world for other governments to do the same thing." Indicative of the then FCC's view of itself, he went on: "For our part if you get a broad market-opening agreement, there is little left that we will need to do in the U.S. to make it a reality. . . . We are very close to the ideal now" (cited in *Communications Week International* 1996, 32). Yet, for the United States, a problem throughout the negotiations was its own market structure, with its division of responsibility between the FCC and state regulators.

In contrast, the E.U. offer made in October 1995 reflected changes within the community. It committed to full liberalization of all basic telecommunications services and infrastructure on January 1, 1998, except for Ireland, Greece, Portugal, and Spain. Those countries had up to four more years to comply. The European Union made its offer and prospective improvement in the foreign ownership restrictions in Belgium, France, Greece, Portugal, and Spain dependent on negotiating partners rolling back restrictions on foreign ownership and assuming a satisfactory level of commitment on an MFN basis—again a reference to the United States' bilateral ECO test.[25]

In a last-ditch attempt to get the negotiations moving, in February 1996 the United States revised its offer to include unrestricted market access and national treatment in the local telecommunications market (regulated by states), and up to 100 percent foreign indirect ownership of common carrier radio licenses. The offer retained a limit on direct ownership but said it was "one of form, not substance" since a non-U.S. company could control a holding company whose subsidiary held the radio license. As previously, the offer was contingent on a "critical mass" of countries providing market access and adherence to regulatory principles.[26]

But there were considerable doubts among other countries as to whether the USTR could deliver on these commitments. The FCC's attempts under the 1996 Telecommunications Act to impose a uniform system of interconnection cost-based charges had already created opposition from state regulators. They had successfully argued in court that local service was a state responsibility and the FCC had no jurisdiction.

The United States Withdraws

The Negotiating Group negotiations finished in April 1996. Only thirty-four members had made schedules of commitments (the fifteen European Union members counting as one) from the thirty-eight negotiating members. Toward the end of 1995, the discussions on how to deal with regulatory principles coalesced into a reference paper drafted primarily by Japan to be used in countries' schedules of commitments where, if included, it became legally binding.[27] Of the thirty-four schedules, thirty included commitments on regulatory principles. But major developing countries such as Indonesia, Malaysia, and South Africa had not made offers, and those from Hong Kong and Singapore limited market access, particularly to satellite services.

Then, responding to the GMPCS operators, Congress (in an election year) became involved in the negotiations. It was concerned that operators in countries refusing access to the GMPCS operators might offer end-to-end services into the United States via other satellite systems—a reversed "Western Union" model. Thomas Bliley, House Commerce Committee Chairman, wrote to Charlene Barshevsky: "We do not want an agreement for an agreement's sake" (cited in Dunne 1996). In other words, the GMPCS operators had decided that no agreement was better than limited market access.

On April 30, 1996, Deputy U.S. Trade Representative Jeff Lang walked out of the talks citing the lack of a "critical mass" of countries prepared to liberalize their markets. The term "critical mass" remained undefined but "included the European Union, Japan, Canada and nearly all OECD countries, as well as nations with highly developed telecom sectors such as Singapore, Korea and Chile, and large developing countries such as Mexico, Brazil, Argentina, India, Indonesia, and Thailand" (Petrazzini 1996, 13).

Following the U.S. pullout, the WTO secretariat moved to extend the deadline for an additional year. The NGBT was to continue its negotiations under the new name of the Group on Basic Telecommunications (GBT). But, before the group could hold its first meeting, acting in response to lobbying by Motorola's Iridium consortium and other GMPCS operators, the United States threatened to exclude satellite services from its WTO offer. In particular, the U.S. move was seen as intended to hamper the privatized offshoot of Inmarsat, ICO Global Communications. The European Union response was that the exclusion of satellite services from MFN was "completely unacceptable." It accused the U.S. satellite industry of blatant protectionism and pointed to similarities with the 1995 financial services talks where, under industry pressure, the United States refused a deal at the last moment (Williams 1996).

Group on Basic Telecommunications

The procedures of the new Group on Basic Telecommunications differed slightly from that of the previous negotiating group in that there were no observers. All who attended were encouraged to negotiate. But the negotiations were affected by the 1996 passage of the U.S. Helms-Burton law. Title III allowed U.S. citizens to take unilateral action against companies trading with Cuba.[28] Mexico complained to NAFTA and the European Union to the WTO. Although in the face of the WTO complaint, President Clinton waived Title III, the act's passage also marked the beginnings of a U.S. trade war on beef and bananas with the European Union.

By October 1996 the U.S. satellite consortia appeared ready to settle and turned to the ITU for help (see chapter 4). There remained, however, the matter of the future of the commercial affiliates of satellite organizations, Intelsat and Inmarsat. Because Washington was concerned that these could gang up with developing countries against the U.S.-owned GMPCS operators, they were left out of the agreement (Aronson 1997, 18). In addition, "accounting rates" were also left on the table (see p. 204).

As the deadline for the end of the talks loomed, Director General of the WTO Renato Ruggiero called on members to make better offers, and the World Bank funded technical sessions for developing country members (Braga 1997, 8n3). Although separately negotiated, the Information Technology Agreement of December 1996, which promised global reductions in tariffs on hardware and software, inspired more liberal telecommunications offers from exporters of electronic products, such as Singapore and South Korea (Friedman 1996). By February 1997 fifty-five offers (fifteen E.U.

members counted as one) had been made, compared with the thirty-four of April 1996. The fifty-five now included the markets of Indonesia, Malaysia, and South Africa and represented more than 90 percent of the $600 billion world telecoms market. Still, U.S. officials described as "very disappointing" Canada's refusal to allow majority foreign ownership of domestic telecommunications companies and criticized ownership restrictions elsewhere (Williams 1997).

Then on February 14, 1997, the day before the end of the negotiations, the United States introduced a GATS Article II (MFN) exemption on one-way satellite transmission of direct-to-home and direct broadcasting services, and of digital audio services. The European Union immediately responded that these were not services covered by the group's negotiations and therefore could not be covered by the exemption.[29] The United States countered that under U.S. regulation these were telecommunications, not broadcasting services, and that GATS did not operate on a single standardized sector classification. The exemption stood.[30] It "allowed the United States complete discretion in choosing which countries could provide direct broadcast services to the U.S. market, a matter of significant importance to Canada" (Aronson 1997, 19–20). In other words, the United States took the MFN exemption as a bargaining chip against Canada's retention of foreign ownership restrictions on its facilities-based suppliers. But U.S. choice of what was "telecommunications" signified a weakness in the basis of the agreement.

The WTO Basic Agreement on Telecommunications was eventually signed in 1997. It did two things: it brought basic telecommunications services under the aegis of the GATS agreement of 1994 in which countries made offers to foreign investors on market access and national treatment (but as U.S. action had indicated, definition of "basic" could differ); second, it produced a reference paper whose provisions could be included within offers. Once included, the regulatory commitments became legally binding and subject to the WTO dispute settlement procedures.

The focus of the final WTO reference paper was on potential abuse of market position by the then state-owned telecommunications operators and the treatment of foreign competitors. Interconnection by new competitors was to be allowed to the dominant operator's own services—in a timely fashion, using "cost-oriented rates that are transparent, reasonable, having regard to economic feasibility, and sufficiently unbundled so that the supplier need not pay for network components or facilities that it does not require for the service to be provided."[31] The wording followed directly the E.U. Open Network Provision directive and was similar to the 1996 U.S. Telecommunications Act.[32] The United States had achieved its "cost-oriented tariffs" for interconnection—both domestic and international.

In addition, there were mechanisms to ensure transparency in the regulatory process, in the award of licenses, and other resources, such as frequency. Developing countries had however succeeded in retaining the right of a member state to define universal service as it wished, but it was to be administered "in a transparent, non-discriminatory and competitively neutral manner" and for new competitors should not be "more burdensome than necessary." These regulatory provisions were to be carried out by a regulatory body "separate from, and not accountable to, any supplier of basic telecommunications services" and were to be "impartial with respect to all market participants."

In effect, the provisions of the reference paper were intended to ensure that network operators could no longer make decisions about the potential entry of competitors. But as Sherman (1998) explains, "In order to take into account different political and legal structures, the paper does not state what entity should carry out the obligations." It could be a semi-independent agency, such as the British Oftel, or it could be a ministry. Nor did the Paper make any reference to U.S. proposals on banning state ownership of operators.

Expanding the WTO/Bypassing the ITU—Accounting Rates

As negotiations reached the final run up to the Basic Telecoms Agreement of April 1997, it appeared that the FCC had been unsuccessful in gaining agreement to make "accounting rates" subject to WTO Dispute Resolution procedures. Reference to "transparency in accounting rates," although in the original draft of the reference paper, was dropped.

Yet "accounting rates" discriminated between countries and were therefore antithetical to both the MFN and transparency provisions of GATS. On the basis that they wished to avoid a future WTO dispute procedure, in February 1997 five developing countries made "accounting rates" subject to Article II MFN exemptions. When others contemplated similar action, the group reached an understanding that "the application of such accounting rates would not give rise to action by members under dispute settlement under the WTO" and that this understanding would be reviewed not later than January 1, 2000, when the next round of WTO negotiations was due to begin.[33] "Accounting rates" had seemingly been left off the WTO agenda by virtue of a gentleman's agreement expressed in a chairman's note.

Developing countries that had not taken a specific MFN exemption on "accounting rates" may have considered themselves covered by this "understanding" of the Group on Basic Telecommunications together with the fact that GATS commitments applied only to those sectors and services specifically included in a member's offer (Tyler 1998, 12). However, those countries that bound themselves to the provisions of the reference paper also bound themselves to "cost-oriented" interconnection. Brazil and India made reservations to the application of the reference paper or future application, but among those accepting it in entirety were not only middle-income countries such as Mexico, South Africa, Indonesia, and transitional economies in Central and Eastern Europe, but also some low-income countries such as Senegal and Guatemala. Perhaps unknowingly they bound themselves to "cost-oriented" international interconnection. The United States was later to argue that the "understanding" did not cover commitments under the reference paper.[34]

During the WTO final negotiations, the FCC had contended that its benchmark determination of 1996 was not extraterritorial legislation. It argued that its refusal of a license for a U.S. carrier to operate a route with a country that did not reduce its accounting rates was simply an "indirect effect," not a unilateral application of domestic law and therefore not contrary to its WTO obligations (FCC 1997). Yet even as the ink was drying on the WTO agreement, the FCC argued that the ITU had been too slow to respond to the issue and determined to impose its benchmark rates (Nye 1997).

The FCC's subsequent 1997 proposals were for even lower benchmarks, together with a schedule that demanded even the poorest countries should comply fully with the benchmarks within five years (FCC 1997, para. 165).[35] Responding to the FCC decision, AT&T argued that the "costs" of termination were universally the same and that "cost-based" rates should not include any contribution to the costs of universal service in the developing country (unlike indirect "costs" as defined by the CCITT). Strangely, AT&T ignored the fact that foreign international operators and those leasing international lines from itself were forced to contribute a 4 percent "universal connectivity charge" so as to subsidize AT&T's universal service to low-income households and high-cost rural areas and access to the Internet. Few questioned the right of the FCC to intervene in private contracts—a repetition of its actions in the 1940s (see Naftel and Spivak 2000, 180–84).

Then a group of foreign operators, led by Cable & Wireless, took the FCC to court on the grounds that it had acted beyond its remit and that it had failed to demonstrate that settlement rates had prevented the fall of U.S. retail tariffs (Scott-Joynt 1997).[36] In late 1997 the European Union Competition Commissioner Karel van Miert weighed in with the argument that all international traffic should be exchanged on the basis of locally determined "cost-based" interconnection (Stern and Kelly 1997, 27). But given AT&T and other carriers' support of the accounting rate system, which provided them with stable revenue, the FCC had no interest in local interconnection. However, the FCC, now chaired by William Kennard, mindful of the isolation of the United States on the issue, suggested to the ITU that it could accept a multilateral solution that achieved "the substantially same result" as the FCC order (Schaffer 1998).

In the meantime the ITU CCITT Study Group III and an ITU Focus Group set up by the 1998 ITU World Telecommunications Forum were unable to reach an agreed solution. Developing countries were unwilling to accept the implications of the growth in bypass of "accounting rates" by end-to-end networks run over leased lines and private international facilities. One estimate put bypass of "accounting rates" at 30 percent of international traffic (ITU 1999a).

Discussion at the subsequent 1998 ITU plenipotentiary became deadlocked. Knowing that its termination "costs" were higher than those of U.S. carriers, a developing-country bloc lobbied for the termination charge (local interconnect) system. But an alliance of countries (led by the United States) that had committed themselves to the WTO reference paper was reported to have "sabotaged attempts to get an early decision." Tactical delay suited their cause. After January 2000 the WTO was empowered to forge a multilateral agreement on accounting rates and the ITU would lose authority to the WTO (Shetty 1998a).

Finally the ITU Focus Group and CCITT Study Group III agreed in June 1999 on a series of "indicative target rates" to be achieved by the end of 2001. The ITU rates differed from those of the FCC in that they proposed lower rates (six cents as against twelve cents) for countries with high telephone penetration and higher rates (forty-five cents as against twenty-three cents) for countries with low penetration. Since the majority of traffic and consumers lived in high-density countries, the ITU rates would have produced a steeper reduction in prices for U.S. consumers than those proposed by the FCC and would have resulted in a more gradual reduction in net settlements in low-income/low-density countries (Utsumi 2000, 7–8). This agreement could have been converted into an ITU-T recommendation by a new fast-track method.

But these rates would also have reduced earnings on the thickest routes for the U.S. operators. The FCC rejected them "on the grounds that . . . adopting these targets would impede progress towards liberalization." This objection was strange if the FCC goal, as it claimed, was the reduction of U.S. consumer charges. But the ITU had found that the averaged differential between the settlement rate on international calls and the billed revenue for U.S. operators had actually increased from 130 percent to 212 percent between 1990 and 1997 (ITU 1999a, 10). In fact the FCC had allowed outgoing traffic from the United States to rise in price (Kelly 1997, 10). The FCC benchmark policy was silent on these price rises and also silent on transit traffic where developing countries wanted lower settlement rates (ITU 1999a, 13). In contrast, FCC rejection of the ITU's proposals was predictable if its aims were to protect U.S. operators and impose liberalization on developing countries—a "Western Union" model.

The United States effectively vetoed the adoption of the ITU-T recommendation until the World Telecommunication Standardization Assembly in 2000 (Utsumi 2000, 7–8). The delay meant that the deadline for the first benchmark implementation passed. Even though Annex E of Recommendation D140 on a transition plan to lower accounting rates was approved by the ITU Assembly, the United States said that it would not implement it (USITUA October 2000, 7).

The FCC action provoked widespread criticism on the grounds that developing countries were cross-subsidizing U.S. companies (see Melody 2000; Thuswaldner 2000). Mark Naftel and Lawrence Spivak (2000, 192n83) suggest there may have even been a diplomatic communiqué on the U.S. contravention of ITU agreements. But in January 1999 the U.S. appellate court upheld the FCC's benchmark policies saying that it was a "valid exercise of the Commission's regulatory authority." Spivak (1999) comments that by giving its approval to FCC actions on the basis that they were taken "to strengthen the bargaining position of domestic telecoms companies in negotiations with their foreign counterparts" and by "condoning U.S. carriers' efforts to engage in what amounts to a group boycott," the D.C. circuit had "bastardized" the public-interest standard of the 1934 and 1996 acts. The FCC now stood for "Facilitating Cartels and Collusion." Although the D.C. court had not ruled that the FCC might lawfully enforce compliance if a foreign government had expressly prohibited such compliance, the FCC subsequently piled unilateral action on unilateral action by enforcement procedures against Kuwait.[37]

However, the FCC also produced incentives to compliance. It authorized international "resale" upon request for any route where the dominant foreign carrier had accepted the benchmark, thereby allowing traffic to be routed outside the settlements process. It also allowed different termination rates at each end of a "resale" route and withdrew demands for proportionate allocation on return traffic. As a result, once a foreign carrier accepted the benchmark rate, with the permission of the FCC it could negotiate "international simple resale" arrangements with U.S. carriers covering 100 percent of traffic on the route. Because the FCC then authorized more than twenty of these routes, it was possible for foreign carriers to reroute public network traffic through a hub into the United States at termination rates that could be as low as four cents per minute. In this way, by bypassing the settlements system, foreign carriers could retain profits while the FCC gained the expansion of international "resale." The FCC also expanded its control over foreign companies via interconnection and licensing: the whole unilateral episode was highly reminiscent of the 1940s (see chapter 1).

In January 1998, only one month after the WTO came into effect, Telmex of Mexico became the first carrier threatened with a WTO complaint for failing to lower its accounting rates (Tricks 1998). Telmex, in alliance with US Sprint, had previously reached an accommodation with the outgoing Hundt FCC in which it had agreed to accounting rates lower than the FCC benchmark rate. But two long-distance competitors to Telmex—Alestra, 49 percent owned by AT&T, and Avantel, 45 percent owned by MCI/Worldcom—capitalized on changed FCC personnel and entered a challenge to that ruling. They wanted local interconnect rates to replace "accounting rates" and an end-to-end network of "resale" over private lines to replace that of the state-to-state network. Mexico's failure was not to have taken a specific MFN derogation on "accounting rates."

In August 2000 the USTR filed a complaint with the WTO. The USTR claimed that Mexico had failed its WTO obligations on three counts. It complained that Mexican domestic interconnection tariffs were high because Telmex had exclusive authority to negotiate interconnection rates for cross-border traffic on behalf of all Mexican carriers—in other words, Mexico outlawed "whipsawing," as the United States had done. USTR complained also that the international accounting rate, although lower than the FCC-mandated benchmark rate, was not "cost-oriented." But it was the third complaint that was the crux of the matter. The two U.S. carriers wanted to be able to provide simple international "resale" that would not only evade the "accounting rate" system and allow interconnection at a local tariff, but would also allow them to collect and deliver traffic in Mexico and undercut Telmex.

Overall, the complaint was a test of whether international interconnection rates (i.e., accounting rates) arranged bilaterally between carriers were subject to the WTO's reference paper provisions on "cost-oriented" interconnection and whether GATS obligations extended to international "resale." Mexico argued that the GATS reference paper related solely to domestic regulation and that "accounting rates" had been deliberately excluded from the WTO treaty by the "understanding" of the Group on Basic Telecommunications. For its part the United States argued that it was the "final text" that had to be interpreted and that FCC benchmark rates were above "cost" based on its long-run incremental cost formulation.

The eventual judgment ran to more than one hundred pages and (contrary to previous WTO practice) was published.[38] The WTO dispute panel concluded that the Basic Telecommunications Group's "understanding" had no legal force. It determined that because accounting rates were interconnection rates for cross-border traffic, and Mexico's accounting rates with the United States were not "cost-oriented," Mexico was in breach of its WTO obligations. It also determined that Mexico's domestic law requiring Telmex to negotiate "accounting rates" and share outgoing traffic proportionately with other operators was contrary to its WTO obligations. In addition, it found against Mexico in that it refused facilities-based operators any interconnection within Mexico, and it refused access to leased lines to commercial agencies located in Mexico. However, despite U.S. protests, it found that Mexico was not under an obligation to allow simple international "resale." In all, the judgment indicated how a WTO dispute resolution was difficult, case-specific, time-consuming, and (given the varying schedules and definitions of WTO, ITU, and domestic legislation) very uncertain.

However, there were implications from the decision—one being that, although the panel accepted there could be a number of methodologies to establish "costs," it adopted a narrow definition of the term "cost-oriented" based on a set of principles regarding how common costs should be allocated to services adopted by the CCITT in 1999.[39] The USTR subsequently identified a number of cases, ranging from high fixed-to-mobile charges in Australia to access charges in India used to finance rural services where it could use the panel's findings (Wellenius, Galarza, and Guermazi 2005, n28). Nevertheless, the judgment did not allow the imposition of international "resale" originally demanded from the WTO by U.S. carriers. The WTO had once again failed to deliver.

After the WTO

The Basic Telecommunications Agreement of 1997 was seen as a major victory for the United States. U.S. private industry was said to have greeted

U.S. negotiators before the end of the negotiations with signs saying "wildly enthusiastic" (Sherman 1998, 15n6). Both President Clinton and Reed Hundt hit the phones to heads of state to ensure that the agreement gathered signatures. The supposition was that the WTO was not only an endorsement of the FCC as a regulatory model but that the market structure of the 1996 U.S. act—demanding "unbundling" of operator facilities and exempting Internet transmission from regulation—was to be developed worldwide. As Hundt said of the WTO in 1997, "By this Agreement, the Telecommunications Act enacted a year ago by Congress has become the world's gold standard for pro-competitive deregulation" (Hundt 1997). To those who may suggest that he was talking to a domestic audience at the time, he repeated the comments when out of office: "Finally we obtained the promises of 69 countries to adopt a pro-competitive policy that mirrored the American rulemaking, including the creation of an FCC-like entity in each country" (Hundt 2000, 203).

According to the FCC, the WTO mandated not only "cost-oriented" interconnection charges but also "costs" calculated in terms of a specific variant of LRIC. There was considerable hubris in the subsequent inability of the FCC to gain nationwide acceptance of its costing of interconnection rates. It was January 1999 before the U.S. Supreme Court reversed some of the findings of the Eighth Circuit Court of Appeals and reaffirmed the nationwide authority of the FCC to set entry prices into the local telephone market (*ITU News* 2/99, 13). But, in turn, the FCC interpretation of the 1996 act, emphasizing low interconnection rates to encourage resellers, failed to recognize as legitimate other regulatory strategies that kept higher interconnection rates to encourage facilities expansion.

The Impact of the WTO

Following the Basic Telecommunications Agreement, there was much press hype about the economic benefits that would accrue to all from "liberalization" of three quarters of world revenues in telecommunications and an assumption that the agreement would result in end-to-end networks. U.S. academic commentators on the overall agreement differed in their optimism. Noam, in particular, predicted a slow, expensive process of dispute settlement and "forum shopping" as governments searched for a regulatory body that would give them what they wanted (Drake and Noam 1997, 814). Other commentators argued that if the idea was to establish centralized, uniform rules, then the reference paper was too vague (Fredebeul-Krein and Freytag 1999).

The multinational users group, INTUG, confirmed in 2001 that, because each country used its own definition of services (as the United States had done for satellite services), it was unclear whether certain "products" (such as financial services) delivered electronically were goods or services. Also it complained that some countries followed U.S. regulation and did not define the Internet as basic telecommunications and therefore it was not subject to the commitments made by member states (INTUG 2001; Naftel and Spivak 2000, 104). However, proposals by the European Union in 1999 to redefine Internet services as basic telecommunications provoked outrage at the proposed overturning of the Computer II decision (Rutkowski 1999).

At the time of the signing of the 1997 agreement, the demand for international telecommunications seemed immense. The Internet was expected to expand demand. Most Internet content and the root servers (the directories holding e-addresses) were located in the United States, increasing international traffic and bringing U.S. operators large transmission profits. Then the dot-com crash of April 2000 ended a week when the U.S. markets lost $2 trillion in value. Followed in 2001 by the attack on the World Trade Center, the resulting fall in demand, overcapacity, and declining margins led to the failure of companies such as Global Crossing Ltd. Its bankruptcy in 2002, after the previous Enron scandal, began an investigation into the links between the company, political donations, and regulatory lobbying (Ackman 2002). In the same year Qwest Communications International Inc. admitted fraudulent accounting, and WorldCom's bankruptcy—at $180 billion, the largest in U.S. history—implicated bankers, the stock exchange, and accounting firms. The seemingly unstoppable U.S. "alternative" carriers had become victims of their own private cables and end-to-end networks.

What of the restructuring of domestic markets? Figures from the ITU in 1999 showed that the proportion of operators with monopoly control over leased lines varied from less than 30 percent in Europe to 50 percent in the Americas, 60 percent in the Asia Pacific, 70 percent in Africa, and 100 percent in the Middle East. Without liberalization of leased lines there could be no international end-to-end networks. In addition, more than 70 percent overall still had a monopoly of basic services. Even in the Americas, only 30 percent allowed competition in national long distance and international services, while the proportion dropped to 24 percent in Asia, 15 percent in Africa, and to less than 5 percent in the Middle East. Only in cellular services and Internet provision had most countries embraced competition (ITU 1999a, 16–18).

As a World Bank paper commented on Asia in 2001, "Somewhat surprisingly, little unilateral liberalization has occurred since the last round of telecommunications negotiations" (Fink, Mattoo, and Rathindran 2001). Yet the bank increased its lending for private investment in telecommunications to $52 billion in 1998—three times that of 1995 (IFC 2004; Winseck 2002). However, in making investments, the bank was held back by the WTO agreement. It could not demand more than compliance with a country's WTO commitment. Coinciding with an internal 1999 report that questioned how much it had helped the poor, there was a redirection in the bank's telecommunications focus toward universal access, rural telecommunications, broadcasting, and infrastructure (see, for instance, Wellenius 2002).

The greatest impact of the WTO was in regulatory agencies. Whereas in the early 1990s there were ten regulators, by the late 1990s there were more than eighty, located in Europe, the Americas, and Africa. Despite U.S. bilateral pressure on some, such as Japan and Taiwan, most Asian countries retained the traditional locus of regulatory power within their ministries. In parallel the developmental states of the Asia Pacific, particularly China, were able to use the mechanisms of privatization, liberalization, and concentration in the sector for industrial policy and societal purposes within the structure of the WTO (Roseman 2005).

From 1997 the USTR reverted to bilateral negotiations and Super 301 threats to achieve its interpretation of the WTO. For instance, a long-running dispute with Japan that lasted into the George W. Bush administration centered on the input costs to LRIC for interconnection charges. A dispute with Taiwan focused on whether it would establish an FCC-type regulatory agency. Complaints from the European Union and Japan against the United States centered on such matters as discriminatory treatment of foreign investment and the abuse of the FCC's licensing power—as, for instance, in FCC action against a Japanese submarine cable and European satellite operators—and congressional action to demand the privatization of Intelsat. Only in the case of the World Com/MCI merger of 1998 were the European Commission and the U.S. Department of Justice seen to collaborate.[40]

After the 1997 WTO agreement, partly in response to E.U. moves to conclude free-trade agreements, under the Clinton and Bush administrations the USTR returned to negotiating bilateral and regional agreements. The first of these with Jordan in October 2000 introduced the term "digital product" into international trade to cover financial services. Further agreements with Chile and Singapore signed by the Bush administration after the 9/11 attack on the World Trade Center were intended to serve as templates for negotiations on the Free Trade Area of the Americas and with ASEAN.

These bilateral agreements went much further in specifying detailed regulatory mechanisms than the WTO. Where the GATS negotiations had used a "bottom up" method, only specifying the services to which commitments extended, the free-trade agreements used a "negative list." Every service was covered other than those specified—so that the agreement would apply to services not yet known. The Singapore and Chile agreements also incorporated many of the demands unsuccessfully made by U.S. corporations prior to the GATS agreement. They included explicit obligations on the provision of "cost-based" access to leased lines (flat-rate and "cost-oriented" in the case of Chile), on number portability, on unbundling of elements of the public network, and on "resale." These bilateral agreements, and that with Australia in 2004, also limited "cultural exemptions" to the traditional audio-visual and broadcasting sectors, ensuring that access for U.S. companies to digitally deliver media over cable, satellite, and the Internet was free from any such national regulatory barriers (Freedman 2006, 29). Together with NAFTA, these "Empire rules"-based end-to-end networks functioned as models for the next round of the WTO.

The New WTO Round

Presciently, writing in 1997, Noam drew attention to the lack of transparency and democracy in WTO proceedings. In doing so, he tapped a vein present in the United States since 1947—distrust of trade policy fashioned by elites. Threatened by a decline in manufacturing and growth of outsourcing, many U.S. commentators felt that the United States had asymmetrically opened its market to global competition and multinational corporations through the WTO (see, for instance, Eckes 1999, 104). From the developing-country perspective there was similar disenchantment. The imbalance of technical expertise between developed countries (Carla Hills at the 1990 Brussels meeting had four hundred advisers) and developing countries, the practice of presenting complicated documents at the eleventh hour and the Green Room negotiations had left many delegates frustrated. Large firms had influenced outcomes. The backlash by public interest groups, labor groups, environmental groups, and developing countries came in the following WTO round, in Seattle in 1999, in Doha in 2001, and finally in collapse of negotiations in Geneva in 2006.

Conclusion

It took fifteen years to negotiate the agreement on basic telecommunications, during which time the world telecommunications scene had altered from universal domestic monopolies to one where the triad of the United States, Japan, and Europe had undertaken liberalization of their domestic markets. The WTO process began with multinational user companies demanding that the U.S. government act in their interests, using the GATT to replace the ITU and Intelsat's regulation of international communications. The idea was to replace the discretion of sovereign governments' over their domestic telecommunications markets with "Empire rules" constructed to replicate the U.S. market through an international organization controlled by the United States. Members of those U.S. user groups were also influential in the E.U.'s initial steps into telecommunications liberalization. They lined up with the European Union and the British against member states' monopoly PTOs. However, the E.U.'s initial hesitancy about negotiations on services allowed developing countries some influence, in that the negotiations on services were conducted separately from those on goods.

The negotiations on telecommunications stretching over ten years illustrated how domestic U.S. regulation was tied to international negotiations, the FCC becoming (during the Clinton administration) almost an offshoot of the USTR. As domestic economic interests gained and lost economic power, so they gained and lost influence in the GATT negotiations. First it was large users and data processing companies that were most influential, then the GMPCS and alternative carriers, then AT&T.

During the GATS negotiations, the USTR was also preparing alternative regional agreements that could be used as leverage within the multinational forum. Similarly, direct relational power against developing countries and withdrawal from the negotiations were bargaining tools. Yet what came out of the telecommunications annex and the Regulatory Reference Paper was not what the United States began negotiations proposing. It did not achieve a system of "top down" rules imposed on national regulators. Had these desires of multinational users been met, in view of the domestic problems of "selling" to Congress even a small international organization, such as the WTO, it would not have had the resources to implement the annual reviews of domestic regulation that the USTR initially proposed.

Instead, developing countries had some influence in the 1994 decisions on GATS. In particular together with the European Union they established a "bottom up" agreement applying only to those services agreed on by in-dividual member states. They also prevented a telecommunications annex that specified "cost-oriented" tariffs. In addition, the regulatory reference pa-per—again, seemingly pushed by the FCC—became benign, predominantly about principles rather than specifying the detailed regulatory mechanisms the United States proposed. Domestic regulators were given discretionary power to regulate the sector according to national priorities.

There is some doubt as to how far developing countries understood that the clause on "cost-oriented" interconnection in the reference paper related to both domestic and international rates—that it would apply to international "accounting rates." In fact, by switching to "cost-oriented" "accounting rates," because of the higher costs of delivery in developing countries it was possible that by the time the U.S. benchmarks were instituted, they could demand higher rates than those of the benchmarks. Some countries even had to raise them to comply.

NAFTA achieved what the WTO did not. It not only enforced the struc-ture of the U.S. market on Canada and Mexico, but it also ensured that those countries' domestic regulators could not regulate VAN operators except in the most limited circumstances. And it instituted the rights of the company over the sovereign state should domestic regulation affect its profitability. It was these provisions, transferred secretly to the OECD's Multilateral Agree-ment on Investment, that were subsequently rejected by France and caused a massive public outcry from NGOs.

Yet the U.S. complaint against Mexico was made under the WTO agree-ment, not NAFTA. NAFTA did not give rights to "resale," whereas the USTR and FCC thought that the WTO did. It may also have had something to do with a new FCC chair determined to make his mark. Yet the complaint demonstrated how the WTO dispute-resolution process could not be relied upon. Although it delivered "cost-based" interconnection to replace bilateral "accounting rates," it did not deliver the international "resale" that the U.S. companies wanted and, by defining the exact meaning of the WTO agreement, may have given greater confidence to watchers from the developing world.

Despite all its hype, as in the 1940s, the FCC was more interested in its domestic power rather than the creation of a world system of "Empire based" end-to-end networks that would have undermined its future role. Unilateral FCC decisions on GMPCS and on "accounting rates," seemingly driven at first by the desire to open world markets, ended in protection of U.S. operators. FCC actions belied its pronouncements. It sought a "Western Union" model of networks, increasing its neomercantilist power. After the WTO agreement, foreign investors into the U.S. market found the FCC a protectionist gatekeeper empowered by its victory over the states and its WTO role.

The FCC interpretation that the WTO reference paper was a worldwide repetition of the 1996 U.S. act influenced U.S. bilateral actions against Japan and others, and ITU interpretations of the WTO agreement. Perhaps because the WTO basic telecommunications agreement had to be sold to Congress as a global liberalizing measure, no U.S. commentator seems to have pointed out how it actually strengthened and centralized regulatory power. To do so would have raised questions concerning its effect on U.S. national sovereignty, something that had hardly been debated.

Yet, starting from a point where the U.S. multinational corporations, particularly those in finance and data processing, had demanded end-to-end networks and the breaking down of national regulation by a system of "Empire rules," the end result of the WTO strengthened the very states that the United States had set out to diminish. WATTC had previously established that end-to-end networks over leased lines could be established subject to rules devised by national regulators. As the FCC and Oftel discovered, even rules that were extraterritorial in application were legal, provided they were nondiscriminatory. The WTO extended the possibility of end-to-end networks to "basic" telecommunications, provided that the other members had not taken an MFN exemption on the mode of delivery. But with the failure of the Multilateral Agreement of Investment, the WTO did little to prevent foreign investment restrictions in many countries, including the United States. Whereas NAFTA weakened state regulation and imposed corporations' rights over sovereign rights, the WTO made state regulation stronger. Countries had only to comply with WTO regulations, not go beyond them. The beneficiaries from the WTO were state regulators. It had delivered a strengthened state-to-state system.

The evidence of WTO failure is in the USTR use of "naming and shaming" tactics to browbeat countries into further concessions on a bilateral basis and its use of "competitive liberalization" to create new bilateral agreements that might spearhead regional bloc agreements for translation into the next WTO. The content of those agreements over the period since the 1997 agreement has been progressively aimed at the denial of discretion to domestic regulatory agencies. Hence, U.S. multinationals and the USTR are girding up for another attempt at "Empire rules"-based end-to-end global networks in a further round of the U.S. "Imperial project."

Conclusion

The story of this book began in World War II, when there were three systems of telecommunications in operation—that of the ITU, that of the United States, and that of the British Empire. We asked what had been the changes in international regulation of telecommunications, where had they come from, who had benefited, and what we could expect in the future. The answers revolve around the linkage between international and national communications networks, between international and national regulation. To help explain what is a complicated story, I developed five potential network models—the "end-to-end" model, the "Empire rules" model, the "Western Union" model, the "European state-to-state" model, and the "WTO model." Only the "European state-to-state" model and the "WTO model" are based on the concept of state sovereignty. This story is about attempts by U.S. agencies throughout the fifty-year period to breach state sovereignty through telecommunications so as to create a U.S. Empire.

In construing international regulation to include unilateral, bilateral, and multilateral actions, our definition takes the concept of international regulation beyond that of the environs of multilateral institutions. I have argued that the concept of "international regime," with its focus on intergovernmental institutions and on consensus, failed to encompass our empirical data detailing conflict and multiple levels of actions. Instead, I have preferred to use concepts of direct and indirect power to explain the empirical data.

The story of the U.S. "Empire project" is a story of disparate attempts by U.S. agencies and personnel to use direct and indirect U.S. power to gain control over sovereign governments. It is the story of how the United States leveraged into international telecommunications its position in the postwar world as the largest, most technologically advanced economy, and its status as military and economic hegemon. It attempted global control through unilateral actions combined with bilateral and multilateral agreements that would establish a system of rules replicating U.S. regulation. These would then be imposed on other governments. A world in its own market image would allow telecommunications to be the electronic conduit for the global expansion of U.S. economic control. It is the argument of the book that in this enterprise, defeated by international alliances and by competition between domestic economic interests and agencies, the United States has only been partially successful.

The book's story lasts from the start of World War II to the end of the century. During that period the "Empire project" went through several phases. This was not an "Empire project" of perpetual confrontation, nor was it fought in one arena. It went in fits and starts as U.S. government agencies gained and lost power either to other agencies or to companies, and as U.S. multinational user companies gained or lost power to U.S. carriers. Congress, in its intervention in trade and in its response to societal concerns for employment, growth, and exports, was crucial to the enterprise. Old and new entrant companies provided the lobbying muscle. New, lower-cost technologies of wireless, satellite, cellular mobile, and the Internet presented the opportunities. The export of domestic regulation provided the means. The rules of intergovernmental international institutions, the creation of private international institutions, the agreements of bilateral negotiations, and the unilateral declarations of domestic regulators provided the platforms.

There were certain periods when, mainly in response to domestic reaction to changes in international economic circumstances, agencies pushed hard against the prevailing international state-to-state system of telecommunications—the 1940s and the New Deal; the 1980s and Reaganism; the 1990s and the neomercantilism of the Clinton administration. There were others when it supported that system—particularly in the 1950s when AT&T was at the height of its power and in the 1990s when AT&T's revenue required protection. In the U.S. political system, the international has always been subordinate to the domestic.

As I said in the introduction to this book, it is necessary in order to explain changes in international regulation to look at the interests of agencies at the domestic level, to look at attempts by domestic agencies to use "policy laundering" (the use of international agreement to leverage domestic change), to look at the shifting of negotiations from one international forum to another, and to look at congressional influence over outcomes. The book demonstrates how the FCC in particular has used action in the international market to counter weakness at home and to placate Congress. Newly appointed agency personnel have tended to be most active—FCC Chairs James Fly, Paul Porter, Alf Sikes, Charles Ferris, Reed Hundt, and William Kennard being some examples. The international is also useful in interagency competition—for instance, the FCC, Comsat, and the USTR have used the international to gain saliency in the domestic arena and in competition with the State Department.

Ideas and their representation of interests have also played their part in the enterprise. After its separation from trade in the 1940s, it was congressional reentry into that issue area in the 1980s that revived the concept of "bilateral reciprocity" and extended it to the telecommunications sector. Only companies from those countries perceived as giving equivalent access to that of the U.S. market would be allowed U.S. entry. The concept allowed the State Department, USTR, and FCC to square the circle of executive commitment to free trade on the one hand and congressional protectionism on the other.

But "bilateral reciprocity" is based on the perception that U.S. markets are more open than those of other countries, on a sense of righteousness that lends itself to unilateral action to protect domestic industry and to control foreign entry. Submarine cable licenses, the 1933 Buy America Act, the Radio Act of 1910, and the "public interest" clause of the 1934 Communications Act could all be used for protection. However, sectoral reciprocity is only useful if there are companies wishing to enter the United States in that sector, and in the early 1980s, when Congress took up the issue, there were only two foreign resale carriers in the U.S. market.

The trade acts of 1974 and 1988, in creating "301" and "Super 301" mechanisms of direct U.S. power, broadened countervailing tariffs to alternative goods and services. They gave leverage to telecommunications and data carriers seeking market entry abroad. In turn, state agencies and companies became closer. USTR complaints concerning market access against foreign countries have been based on information from companies for which corporations have not been publicly accountable. As the agency and U.S. companies have become mutually dependent, so neomercantilism has become the rule—expansion abroad, protection at home.

Other ideas have also created the base for U.S. actions. The "free flow of information" concept of the 1940s legitimated neomercantilist actions by the FCC. "Public choice" theories of the 1970s and 1980s legitimated "privatization" as a policy of the Reagan and Thatcher administrations to be pressed on developing countries through the World Bank. The "Washington consensus" of the 1980s and 1990s led to market restructuring and the opening to Western investment of those countries that faced financial deficits—except for the United States. Fear of foreign nationalization led to investment guarantees and (much later) to demands for rights to investment for multinationals. And in the regulation of telecommunications, as the first to open up a monopoly, the conceptual structure adopted by the United States became a global standard. That standard had become so inherent that the FCC perceived the WTO agreement of 1997 as world acceptance of the 1996 U.S. act—and took commensurate action when that failed to be the case.

Domestic lobbying from industry, channeled through Congress, has been at the root of U.S. actions. In a political system that requires large donations from industry, lobbies have risen and fallen according to political and/or economic importance. In the 1940s AT&T and the international record companies formed the strongest lobby. In the 1950s AT&T's control of coaxial cables made it hegemonic. But from the 1970s, as domestic liberalization began, large-user companies gathered strength. The small group of companies such as Citibank, American Express Co., and IBM Corporation, which pushed in the early 1980s through Congress for the introduction of telecommunications into the General Agreement on Trade and Tariffs (GATT), intended that U.S. companies should gain rights of investment overruling national sovereignty and that an agreement should provide worldwide enforcement of rules to ensure end-to-end private networks. Then it was the aerospace satellite manufacturers and new entrants, such as Panamsat, who pressed for the liberalization of Intelsat's maturing market in the 1980s. It was data processing and information service suppliers who pushed for the opening of the international market to VANs in the 1980s. It was the electronics industry who supported the election of President Clinton and who were the anticipated beneficiaries from the global information infrastructure with its promise of worldwide private networks. Global personal communications satellite operators had strong congressional support as they set out to breach foreign PTO monopolies, demanded end-to-end international public networks and protection from Intelsat and European operators. Then, in the negotiation of the 1997 WTO agreement, large users, traditionally allied to the USTR, eventually lost out to AT&T in alliance with an FCC led by someone close to the Clinton administration. Subsequently, carriers also developed close relations with the USTR in the transmission of complaints against foreign countries. Hence the book illustrates how the U.S. domestic political system has created the rise and decline in influence of U.S. corporate lobbies on the "Empire project" throughout the period.

The book demonstrates that a traditional response to domestic economic pressure has been a unilateral edict from the FCC or from the administration. Such unilateral U.S. actions are part of the fabric of this story from the 1940s through to the 1990s. They include FCC actions on tariffs in the 1940s; congressional creation of Comsat in the 1960s; FCC decisions on VANs of the 1980s; FCC licensing of private submarine cables in the late 1980s; administration decisions on Domsat and Intelsat in the 1980s; FCC licensing of global mobile communications satellites and FCC unilateral decisions on "accounting rates" in the 1990s.

A second response, at the heart of the "Empire project," has been an attempt at the control, replacement, bypass, or restructuring of multilateral institutions—particularly those based on some form of internal democracy. Institutions outside the U.N. family, such as Intelsat, the World Bank, the OECD, and GATT, have allowed the United States to exercise more direct and indirect power than has the ITU. Intelsat began life under direct control of the FCC, then, at U.S. insistence, became an international organization with weak internal democracy. The U.S. president and the U.S. treasury control the appointment of the World Bank president and at times have bypassed even the limited democracy of formal internal procedures to gain bank action. OECD has had a very limited membership, and its potential for secrecy provided the forum for the Multilateral Agreement on Investment. GATT was attractive because it had a dispute-resolution mechanism, conducted by appointed personnel, nonaccountable to any democratic process.

I suggested in the introduction to this book that the United States has disliked U.N.-based international institutions. What is recorded here concerns how U.S.-led restructuring has helped to mitigate the power that the one nation–one vote system gives to the majority. Restructuring of the ITU, both in the 1940s and the 1990s, was intended to increase U.S. indirect power through its industry's increased influence—until industry autonomy threatened FCC power as a gatekeeper. Failing to control or replace the organization, the FCC then forum-swapped—as with "accounting rates"—or bypassed it with unilateral actions.

A third response to domestic pressure, particularly where the United States has had an initial monopoly of a technology, has taken the desire for nonaccountability further into the creation of private corporations linked to defense agencies and under control of favored elites. Comsat in the 1960s and ICANN from the 1990s gave the United States direct power to govern new technologies on a global basis and to bypass other governments. ICANN represents the summit of this process—private, nonaccountable, nondemocratic, and under direct U.S. control.

And finally, where the United States has failed to gain the "Empire rules" system from multilateral agreement—as in the case of the WTO—it has followed through with bilateral agreements. These, like NAFTA, impose the rights of corporations to foreign investment and limit the rights of domestic regulators. Such agreements are "Empire rules" based, overriding respect for national sovereignty and, ironically, subjugating U.S. court decisions to international, nonaccountable tribunals.

It is the U.S. attempt over fifty years to impose such "Empire rules" worldwide in the interests of its corporations that is this story. The background is the organization of European international telecommunications as a state-to-state model with interconnection at the borders of each sovereign state. Arranged in the 1860s to allow the international transmission of telegraphs, there were two major exceptions to this state-to-state–based system. The one was that of the British Empire, which brought together a range of Dominions and colonies in Africa, Asia, and the Caribbean to act as one end-to-end unit under the international monopoly of Cable & Wireless. The second was the U.S. network controlled by U.S. international record companies under the regulation of the FCC.

The two systems had different origins. The British Empire network came out of the decision in 1927 by the British government to protect its submarine cable network from the lower cost infrastructure of shortwave wireless. When the company showed signs of failing in 1937, the British government extracted from the Dominions the promise that they would not set up competitive wireless routes. In return, those Dominions received a weak power of regulation over the company's tariffs. For its part, under the "special arrangements" clause of the ITU regulations, the company instituted a preferential flat-rate British Empire tariff and a very low flat-rate tariff for the British press. The British Empire system, still formally part of the ITU state-to-state system, constituted a barrier to U.S. companies seeking worldwide operation and to U.S. content producers.

In contrast, the U.S. network was never part of the ITU system. From the 1860s and the first transatlantic submarine cable, overseas operation was an extension of the national network. Whereas the British company, Anglo-American, relied on Western Union to deliver its traffic within the United States, U.S. companies collected and delivered traffic within the United Kingdom. Then, in 1911, when AT&T/Western Union refused interconnection to the British, the transatlantic "Western Union" model came into existence, to last until the 1960s. However, when U.S. radiotelegraph companies began operation in the 1920s, they did so on the basis of the state-to-state system, as did AT&T for the radiotelephone. There was no domestic regulation of international telecommunications under the Inter-State Commerce Commission from 1910 or in the early years of the FCC. And, because U.S. membership of the ITU would legitimize domestic regulation, the companies successfully opposed the U.S. government's accession to the treaty. In particular, AT&T marketed its role in the domestic economy as a "public interest" monopoly.

The first phase of the postwar U.S. "Empire project" came in the 1940s. James Fly, an ardent New Dealer, used FCC wartime powers to take control over the international record companies and displayed the FCC's relational power by enforcing reductions in tariffs, first where both ends of an international route were owned by a U.S. company, and then on the state-to-state systems of the radiotelegraph companies. Those governments whose rights were overridden by the FCC included France and Italy (liberated by U.S. troops) and South American governments. Using its indirect, institutional power over interconnection, the FCC also expected Cable & Wireless to carry out its orders.

But the FCC unilateral actions to create an international system under FCC control came up against the British Empire system. Bilateral agreements seemed to effect a compromise but were no sooner made than broken. In effect, international intergovernmental relations were secondary to the domestic politics of the New Deal. The British and the U.S. State Department then formed an alliance to bring the FCC under control.

Chairman Fly himself envisaged a utopia in which everyone in the United States would be able to communicate by telegraph throughout the world at low, uniform rates similar to those of the postal system. His rhetoric against the British Empire system and propaganda on the virtues of the "free flow of information" became more strident as he suffered at the hands of Congress. With the intention that the previous British hegemony of submarine cable communications should be broken, Fly proposed a merger of all the U.S. international carriers.

In the face of objections from other agencies, he was only able to persuade Congress to agree to the merger of the two domestic companies. But he then used that merger to extend FCC domestic power. Inadvertently, Fly created an almost complete division (AT&T being the exception) between "national" and "international" networks, reinforcing a state-to-state system. He then cartelized outgoing traffic into the "international formula" that brought the companies under the FCC until the 1980s, when the agency similarly extended its power over competing telephone companies. It was the impact of this formula on company behavior that was at the root of the "accounting rates" crisis in the 1990s.

Liberalizing the British Empire monopoly was a goal for both Fly and the radio-based U.S. carriers. U.S. entry into the war gave the opportunity. Having always shared traffic with their opposite numbers overseas, now radio companies wanted rights to carry third-party traffic over their public networks, thereby creating a U.S.-dominated global network. Even AT&T, which had always worked on a state-to-state basis, attempted to set up a company in British-dominated Egypt for an end-to-end network. But U.S. entry to the British Empire did not yield the worldwide liberalization initially sought. The British used the technique of requiring commonwealth agreement in order to limit the postwar retention of direct lines and to prevent the U.S. radio companies taking third-party traffic.

At the same time, in defense of its Empire network, the British government was obliged to allow more autonomy to the Dominions, and used the postwar threat of competition to nationalize Cable & Wireless. It transferred the company's overseas assets to Dominion governments. Subsequently, as colonies became independent, they bought out Cable & Wireless's local assets and created monopoly PTOs. The commonwealth was to remain a regulatory bloc, with its own system of accounting for the transmission of traffic, until the 1980s. Its "wayleave" system eventually fell apart because of British entry into the European Union and because its central regulatory panel could not regulate domestic systems against the wishes of increased numbers of sovereign members—exactly the international regulatory failure that U.S. administrations later set out to rectify through the WTO. However, particularly up to the 1970s, the commonwealth bloc was to thwart U.S. Empire-making enterprise.

The second phase of the U.S. "Empire project" came at the end of the war and involved the defense and expansion of the protectionist "Western Union" model of international networks and the restructuring of the ITU. Initially, U.S. State Department intention was to replace the ITU with an organization under direct Allied control. As with the failed International Trade Organization, it wanted a body with supranational authority. Then it sought to bring the ITU under the direct control of the United Nations, an organization then subordinate to the United States' political hegemony, and to transfer the ITU from Berne to New York. The British Post Office, fearful of the influence that the ITU's location in the United States would give to U.S. manufacturers, together with allies in Europe and the commonwealth, thwarted these plans.

Why the United States agreed to become a member of the ITU is unclear from the papers, but it may have been part of the FCC's domestic campaign to establish itself over the international record companies. Membership of the ITU strengthened the FCC. It became the representative of companies' interests in the ITU and controlled whether operators and their manufacturers could attend CCI meetings. Subsequently, the FCC established the right for U.S. carriers' public telegraph end-to-end networks to bypass the ITU's international regulation through the "special arrangements" clause in its convention. The U.S. "Western Union" model network thereby gained the same international legitimacy as the British Empire system.

Yet AT&T was so powerful that its interests were sufficient to ensure that the United States did not abide by the international treaty that obliged its signatories to sign the telephone regulations. Whereas the United States signed the telegraph regulations in 1950, it did not sign those for the telephone until 1973—and even then they excluded networks with Canada and Mexico. Weakened by congressional action, from the 1940s until the 1970s, when the White House began intervention, the FCC was less powerful than AT&T.

From the 1950s the development of coaxial cables further strengthened AT&T's domestic and international monopoly and the state-to-state system. For the British, the state-to-state basis of coaxial cable, which could carry both telephone and telegraph, provided the possibility of ridding Britain of the "Western Union" system of the U.S. international record companies. Although the FCC, under pressure from the record companies, would not allow the new cables to carry both telegraph and telephone, and ITT attempted to bypass the "international formula" with its own coaxial cable, through TAT-1 the British eventually succeeded in gaining the state-to-state system they wanted.

With the development of transatlantic coaxial cables and the growth of postwar nationalism the end-to-end, U.S.-controlled international public telegraph networks faded. In addition, the commonwealth hit back against FCC protectionism. Instead of taking a share in AT&T's TAT-2 cable, the British and Canadians laid the first commonwealth cable, CANTAT-1, which bypassed FCC regulation and carried both telegraph and telephone. The FCC was forced to allow the U.S. international record companies to lease lines on the TAT and CANTAT cables and then to own them. Such ownership turned them into state-to-state–based companies and from the 1960s their transatlantic "Western Union" network ended.

The rise of postwar nationalism in South America also spelled ITT's demise as a foreign domestic operator. Following the company's involvement in the toppling of President Allende in Chile, ITT's withdrawal from South America strengthened the state-to-state system. Suspicion concerning the company's part in the military dictatorship in Chile helped fuel Latin American support for the autarky of "dependency" theory in the 1970s. Later, the ghost of ITT's experience with nationalization and compensation returned in the 1980s in the form of demands by multinational corporations for supranational invest-ment rights to fuel another phase of the "Empire project."

During the decade from the early 1950s, it appeared that the state-to-state system would become hegemonic. However, following the success of the CANTAT-1 cable between Canada and Great Britain, the British Common-wealth decided to renew the old global Cable & Wireless network with co-axial cables and excluded AT&T. The idea was for a renewed British Empire system. Similarly, when the United States gained the monopoly over satellite technology, it tried to prevent the Europeans benefiting.

Fuelled by U.S. nationalism, commercial satellites provided the opportu-nity once more for U.S.-controlled global networks. In the third phase of the U.S. "Empire project," the private U.S. corporation Comsat tried to establish end-to-end public telephone networks. Despite the Kennedy administration's intention for satellite technology to be universally shared, and despite oppo-sition from senators and trade unions in favor of a public-controlled entity, Congress created Comsat. The first chairmen of Comsat were interested only in unilateral or bilateral arrangements in which Comsat would provide an end-to-end service via earth stations at home and abroad. The State De-partment worked to place Comsat personnel into uncomfortable positions in their negotiations with the Europeans. British pressure on the divided Europeans brought eventual acceptance by Comsat of joint ownership of Intelsat and an internationalized board of governors. But at U.S. insistence Comsat remained manager, the United States retained a veto, and Intelsat had a monopoly of satellite carriage. Lacking the technology, the Europeans had little bargaining power.

Intelsat was an international organization in name only. Until 1973 it had no legal identity and was subject to direct FCC regulation under congres-sional control. U.S. satellite manufacturers benefited. As both the manager of the system and U.S. representative on Intelsat's board of governors, Comsat was able to change hats at will. It was this conflict of interest and European perceptions that Comsat's behavior was an abuse of U.S. power that welded divergent views in Europe into an overall aim—to get rid of Comsat as man-ager and, based on French ambitions, to develop Europe's own technology.

Unwilling to lose its monopoly position, Comsat attempted to prevent the conclusion of the Intelsat definitive arrangements and its replacement as manager. Because U.S. domestic legislation named Comsat as the U.S. signatory, the State Department would not allow Intelsat to become an intergovernmental U.N.-type institution. From 1973 the United States insisted on an Intelsat structure along the lines of a monopoly European PTO, responsible for both operation and regulation, thereby benefiting its satellite industry. At European insistence, however, the final treaty allowed a potential breach of Intelsat's monopoly through the clause in its convention that allowed regional satellite systems that did it no economic harm. With the U.S. veto abolished, the final treaty was perceived as a disaster for U.S. interests. Yet U.S. commercial interests and Comsat were still able to use institutional power to hinder European satellite systems through interpretation of the Intelsat definitive agreements.

AT&T and the record carriers prevented Comsat from providing end-to-end service by imposing joint ownership of earth stations. Intelsat then reinforced the state-to-state system using national operators as its gatekeepers to domestic networks. Yet, although Intelsat was primarily run for their benefit, European state-owned operators were unwilling to leave their destiny in the hands of an organization over which they had little control. Their interests married with those of AT&T and the international record carriers, dependent on investment in submarine cables to inflate their rate bases.

AT&T was able to utilize the support of the record companies and of the European PTOs to gain licenses from the FCC for submarine coaxial cables. TAT followed TAT, each with more capacity. A weak agency, unable to regulate AT&T, the FCC failed to ensure that the cable companies filled the existing capacity on Intelsat's satellites, which remained half empty. By virtue of U.S. operation of the eastern end of the most-used and profitable international route, the FCC became an unacknowledged global international regulator, able both directly and indirectly to govern Intelsat through transatlantic capacity.

Only in 1971 and 1975 did the FCC attempt to stand up to AT&T. Then the combined onslaught of criticism from European PTOs and AT&T produced a congressional response in favor of limitations on FCC power. Despite formal mechanisms of transatlantic cooperation, the FCC became a unilateral regulator, its domestic weakness ensuring that Intelsat could not compete directly with AT&T for the ten years it took for the new technology of optic fiber to come on stream. That technology then swung cost-effectiveness for long-distance point-to-point communications away from satellites. Developing-country clients of Intelsat and U.S. consumers who financed much of the submarine cables were the losers from FCC global control.

Only when it became evident that the Europeans had their own satellite technology and were intent on separate systems did a U.S. domestic satellite become politically acceptable. Led by the Nixon White House, the unilateral policy of "open skies" was intended to expand exports of cultural products to Europe and give U.S. satellite manufacturers a larger market. This was the time when computing and telecommunications began to converge and digital switches became the new technology. IBM's entry into data transmission in competition with AT&T came through the Domsat policy, the rationale for which envisaged data transmission between large companies free from regulation.

This Nixon White House gift to IBM was the forerunner of the FCC's 1980 Computer II separation of "enhanced" from "basic" services. It was the "open skies" policy that reinforced the trend to company-to-company, end-to-end domestic data transmission and to the beginnings of international end-to-end "enhanced" services or "Value Added Networks." Through the local efforts of IBM the Computer II regulatory division was exported to Britain and Japan, leading to the next stage of the "Empire project."

Faced with demands from large users for a more integrated domestic/international service, FCC Chair Charles Ferris once more allowed the international record carriers to compete in the domestic market and breached the divide between domestic and international. But its unilateral decision to expand the domestic nonregulation of "enhanced" service providers and domestic "resale" into the international arena brought it up against the ITU and CCITT regulations that supported the state-to-state system. Bowing to AT&T's fear of European retribution and ITU opposition, the FCC drew back from that proposed liberalization.

Soon after, the Reagan government used the "enhanced" and "basic" divide to end the role of Intelsat as regulator of the global satellite market. That the U.S. government felt it within its power to reinterpret unilaterally an international treaty demonstrated how Intelsat's geographical location weakened it. Intelsat's position was also undermined by the fraudulent behavior of its chief executive and the imposition of his replacement by the White House. Despite passive opposition by Intelsat's members to the new entrants, by the late 1980s Intelsat was a regulator in name only. Then, when optic fiber transatlantic cables came on stream, the FCC bypassed ITU regulations by unilaterally licensing private entrants. Intelsat began to suffer from declining revenue. For the first time its vulnerability became evident and its eventual privatization became almost inevitable.

Having broken Intelsat's monopoly, the fourth phase of the "Empire project" started in the late 1980s, with attempts to bypass ITU international regulations. Privileged by the Computer II division, large users and equipment exporters sought the international replication of the U.S. domestic regulatory model. "Future proofing" of their foreign investment meant ensuring that domestic regulators in foreign countries could not regulate private end-to-end networks. They wanted rights to investment, to lease lines at bulk cost-oriented rates, to interconnect into public networks at cost, to use proprietal standards, to sell to third parties, and to use private infrastructure as they wished. They sought an "Empire rules" system that would prevent any discretion in foreign regulation of their networks by a system of rules that would be created by the United States and imposed supranationally by a dispute-resolution panel. On the assumption that the U.S. market was already "free," the impact of such rules on the United States itself was never debated.

In this phase of the "Empire project" numerous attempts were made to bypass, alter, or replace existing international rules of regulation, to alter the balance of power in institutions that created those rules, and to alter systems of interconnection and payments that supported the state-to-state network model. All forms of mechanisms were used—unilateral, bilateral, and multilateral—with the intention that one level might be used as a lever in another, and all forms of power, both direct and indirect, were brought into play.

The first multilateral forum to feel this pressure was the ITU. WATTC-88 provided the focus. In the intervening time, domestic liberalization had taken place in the United States, Japan, and the United Kingdom, and the European Commission had produced its green paper that set out to harmonize the E.U. market. During the 1980s, U.S. companies, such as IBM and GEISCO, had been successful in achieving international value-added networks with Japan. By following U.S. domestic regulation, the designation of VAN suppliers as recognized private operating agencies enabled them to bypass ITU regulations. In effect, these networks transferred the U.S. domestic division of "basic" and "enhanced" into the international arena.

But European PTOs, particularly those in southern Europe, saw WATTC-88 as a potential forum to bring these "enhanced" or "value added" services firmly back under ITU regulation. In particular, France could see no reason why publicly owned data networks should be opened up to private U.S. companies. State-based operators and AT&T dominated the preparatory work for the conference. The U.S. representative from the State Department signed off on the resulting reference paper. But the new U.S. entrants produced domestic opposition to the paper that spiraled into lobbying of the ITU and all governments over which the U.S. held direct or indirect power.

The ITU secretariat had the unenviable task of attempting to please its largest donor while not alienating the mass of developing countries that formed the majority of its membership. Acting outside its normal remit, the secretariat supported the liberalizers (led by the United States), manipulated the conference, and succeeded in alienating both constituencies. Driven by new entrants, against the advice of AT&T, the United States proposed regulations that removed the stated right of countries to regulate "enhanced services" networks. It provoked a reaction from African administrations and an eventual compromise that recognized the status quo. U.S. refusal to compromise led to political isolation. Later, in negotiations within the CCITT to alter its recommendations in line with the new WATTC regulations, it was not U.S. power but the fact that the Treaty of Rome had taken precedence over the ITU regulations and recommendations that determined the eventual outcome.

Following WATTC-88, the ITU's "special arrangements," whose definition the U.S. government had demanded in 1949, now included international "resale" and allowed "enhanced services" between consenting members. Suppliers of these services could use proprietary standards and fail to interconnect with each other. The U.S. domestic market had thus been exported into international regulations.

But the tradeoff in the WATTC package was the specific recognition that governments had the sovereign right to decide on the structure and regulation of their domestic networks. WATTC-88 not only left the state-to-state system intact, it strengthened it. Subsequent domestic regulation in the U.K. case demonstrated that WATTC allowed extraterritorial regulation so as to control the overseas behavior of the operators of end-to-end private and public networks. In the U.S. case, it allowed regulation based on "bilateral reciprocity" that refused entry to operators from countries of whose domestic regulation the FCC disapproved.

From the perspective of the U.S. companies and their liberalizing allies in Europe, WATTC-88 demonstrated all that was wrong with the ITU. In particular by the late 1980s the ITU had moved from an engineering club of the 1950s to numerical domination by developing countries. Well before WATTC, the ITU plenipotentiary meetings reflected the tension between the traditional standardization focus of the ITU and the development problems of the majority of its membership.

In the 1970s and the early 1980s the numerical superiority of the developing countries and the impetus of OPEC's display of power allowed them to gain some change within the established practices of the ITU. But U.S. delegations were hostile. They used procedural and financial maneuvers to ensure that the developing countries did not gain a development section financed from the ITU's budget. The primary conflict within the organization was north vs. south, but unlike UNESCO, the United States and the Soviet Union were in the same coalition. In the early 1980s, U.S. agencies contemplated withdrawal from the ITU or restructuring it.

Then the debt crisis of the 1980s undermined the position of developing countries. In the late 1980s the threat to the ITU central role as a standardization body came from U.S. attempts to replace its standardization functions with those of alternative regional agencies. The prospect of the ITU's becoming irrelevant to its major donors coincided with an imminent change in leadership. Hence, restructuring instigated by ITU leaders prevented U.S.-inspired bypass of the organization.

Restructured into three sections in 1991, in which the developing country members obtained their own section with its own plenipotentiary, the changes were presented as progress for them. But the restructuring effectively sidestepped the one nation–one vote basis of the organization. As first mooted in the United States in the early 1980s, the restructuring insulated the standardization functions of the industrialized and left the developing countries talking to themselves. Direct financial pressure from the United States and the United Kingdom meant business was given more influence within the organization, and a change of leadership coincided with a change in ITU policy. Internal discussion began on how private industry could gain votes on a par with governments.

Whereas previously the ITU had espoused the developmental role of the state, its personnel now fell in with the U.S.-led "Washington consensus" and promulgated the supremacy of markets and loss of state sovereignty. That change did not protect the organization from U.S. and E.U. attempts to bypass it through the creation of a WTO committee to monitor and enforce implementation of domestic liberalization. In the ongoing debate on the role of the ITU, the World Bank became influential within the organization. Bank telecommunications personnel gave it direction by espousal of domestic regulation as a counterpoint to privatization. Hence, in 1993 the ITU held its first forum on regulation—one year before the creation of the WTO and four years before the mainstream World Bank acknowledged that the state had a role in the structuring and overseeing of the market.

The change of emphasis within the ITU coincided with changes in the world economy. For developing countries the demise of the Soviet Union in 1990 had removed the benefits of neutrality between East and West. After the first Iraq war, the United States was the military and economic hegemon. Recession in the West decreased aid so that developing countries had to rely on foreign investment for capital. Multilateral financial agencies, the IMF, the World Bank, and the IFC gained power, as did multinational users and operators looking for overseas investment to offset monopoly profits at home. U.S. propaganda, through the global information infrastructure, was directed to ensuring that foreign domestic markets were opened up to U.S. investment.

During the 1980s the World Bank had become the initial regulator of the telecommunications markets of those who borrowed from it, determining their structure. U.S. power over the bank was both direct and indirect, through finance and personnel and lobbying from Congress and presidents. The Bank's geographical location weakened its independence. The Reagan administration successfully pressured the bank to prioritize the sale of state assets, and the first Bush administration attempted to have half the bank's resources go to private industry. Particularly after the fall of the Soviet Union, the IFC, with its loans to the private sector, gained in influence. The beneficiaries were Western, particularly U.S. companies.

Although it began telecommunications lending in the 1960s on a project basis, it was after the introduction of structural adjustment lending in the 1980s that the World Bank's minor financial input into telecommunications enabled it to act as the facilitator and enforcer of loans from other sources. In turn, because it had to convince those other entities to loan money, its policy of privatization had to be seen to be "right" and the output of bank-funded research, consultancies, conferences, and publications all followed the bank's line.

But the World Bank telecommunications personnel pursued a somewhat different agenda from the mainstream World Bank. Even before the privatization of BT in 1982 and the establishment of Oftel, they argued for the autonomy of telecommunications from government bureaucracy and for regulation of operators. However, early attempts to establish a regulatory agency in Argentina made it evident that the role of the World Bank as a lending institution did not allow for the ongoing involvement that regulation in a developing country required. Through the ITU, despite failures (such as the green paper on Africa), the bank's telecommunications personnel were able to promulgate the doctrine of privatization and regulation, roughly based on the British model.

The first bank-led privatizations, for the benefit of U.S. commercial banks, were later criticized for providing windfall profits to the buyers. Public opposition prevented privatization of telecommunications in a number of countries. And regulation, by preventing the opportunities for rampant private monopoly, could also prevent a sale. For richer developing countries it was possible to follow the model of British privatization with a partial international public offering, but for poorer developing countries where outright sale of the PTO was too politically sensitive, cellular technology provided an alternative.

License fees from the introduction of cellular operations produced cash resources and placated the World Bank. The outcome, although presented by bank personnel as "privatization," was actually competition. By 1998, half of the countries in Africa had allowed competition in mobile, while 90 percent still had a PTO monopoly of local, long distance, and international service, and fewer than ten countries had privatized the operator. But the award of cellular licenses then provided the opportunity for the exercise of U.S. and European structural power over previous colonies in favor of "their" equipment standards. European GSM for second-generation cellular technology became the global standard except where the United States had political influence—for instance, in Russia, Central and Eastern Europe, and the Americas.

The fall of the Soviet Union in 1990 provided the opportunity for the World Bank to become involved in the previous Soviet Central European colonies in what were telecommunications networks suffering from decades of neglect. U.S. companies, the World Bank, and the IFC were quickly into the new market. The World Bank now had a competitor. But because the European Bank for Reconstruction and Development would not allow manufacturers to form part of the consortia it backed, the World Bank's conditions were more favorable to equipment exports. In Hungary, the World Bank's decision to make the telecommunications market into an experimental copy of the system in the U.S. Midwest of the early twentieth century led it into conflict with other advisers. In Russia, the sale of state assets through a voucher scheme in a "Big Bang" of privatization proved a failure. Privatization proved not to be a panacea.

Yet the IFC continued to advocate privatization without regulation. Had it not been for the World Bank's telecommunications staff's pressure to initiate domestic regulation in developing countries, the bank (and particularly the IFC) would have been little more than an overt facilitator of the crudest economic imperialism. The sale of operators to Western interests once more opened the door to end-to-end colonial, privately owned, public networks. Yet, overall, the impact of the World Bank by the early 1990s had been to strengthen the state-to-state system, reinforcing private monopolies and reintroducing old colonial ties.

However, strengthening the state was no part of the mainstream bank's privatization agenda. It was not until 1997, three years after the inauguration of the WTO, and in the same year that the basic telecommunications agreement was signed, that the World Bank relented. Having excoriated state-based institutions for the previous fifteen years, it began to acknowledge that in order to regulate, there had to be a state administration capable of doing so. By then, cellular operators, among others, were demanding regulation of interconnection in particular.

Into the bank-supported international state-to-state system (now increasingly interlinked by European GSM cellular technology) came the GMPCS operators of the 1990s and a new phase of the "Empire project." Consisting primarily of U.S. companies, with technology developed from the military, the FCC unilaterally gave them international licenses. Once again, these U.S. operators intended end-to-end networks to break open the state-to-state system. But the lessons of Comsat had been forgotten. Despite unilateral licensing, operators could not establish end-to-end networks without cooperation from the national governments whose territory they intended to breach. Originally aimed at the PTO monopolies of Europe, by the time of their launch it was Asia, with its undeveloped cellular market, that they hoped to fill.

But Asian governments were suspicious of the intentions of these U.S. companies. To some extent the ITU proved useful to the U.S. corporations and their government in negotiating a memorandum of understanding between the companies and foreign governments, but the GMPCS technology was too expensive compared to cellular technology and ended in bankruptcy or use by the U.S. military. This further attempt at end-to-end networks on the part of the United States was not only a failure, it also led to major problems with the European Union when, in another example of the "Western Union" model, the FCC discriminated in its licenses against the one European operator.

As the GMPCS venture was failing, a further attempt at the United States' "Empire project" was also failing in the Basic Telecommunications Agreement of 1997. The negotiating process had begun in 1985 when only the British and Japanese markets were liberalized. U.S. multinational user companies had demanded their government act in their interests to force the opening of domestic markets by use of international rules and dispute resolution mechanisms. In GATT they saw an international organization that could be used to combat the European-dominated ITU and Intelsat and those institutions' support for the state-to-state system.

The European agreement to enter talks on services had much to do with a tradeoff against liberalization in agriculture. Later the negotiations reflected, rather than led, the ongoing internal E.U. liberalization of telecommunications networks. Despite direct U.S. arm-twisting based on economic structural power, developing countries had some influence in the 1994 decisions on GATS. In alliance with the European Union they gained separate negotiations of goods from services. They gained the withdrawal from the WTO discussions of multinationals' rights to investment and gained a telecommunications annex that allowed them "get out" clauses, recognized the ITU, and failed to specify "cost-oriented" tariffs. They also eventually established a "bottom up" agreement applying only to those services agreed to by individual member states.

The United States introduced proposals into the GATS negotiations under its 1990 telecommunications annex that would have introduced an "Empire rules" system. The U.S. annex anticipated enforcing domestic liberalization so that other governments would have no discretion in the implementation of U.S.-determined rules. Those rules would have created rights of companies over governments. Even while it was negotiating in the WTO, the USTR was also preparing alternative regional agreements that could be used as leverage within the multinational forum.

When its telecommunications annex was bypassed within the WTO negotiations, the USTR transposed it into the NAFTA negotiations between the United States, Canada, and Mexico. NAFTA placed corporations above governments, allowing them to sue with regard to public policy detrimental to their profits. It transferred power over binding decisions to nonaccountable elites. It was an "Empire rules" model defeated in the GATT/WTO negotiations because of developing-country alliances and their use of the tradition of GATT consensual decision-taking to counter U.S. power.

NAFTA introduced fierce constraints on what national regulators might demand of value added network operators. Companies gained the option to use private networks for whatever they wanted. NAFTA was a success for U.S. multinationals. They wanted the same "Empire rules" in the WTO.

But, in fact, the 1994 GATS agreement and telecommunications annex produced little liberalization of domestic markets. Yet, even had the desires of multinational users for an "Empire rules" system been met, in view of the domestic problems of selling to Congress even a small WTO organization, it would not have had the resources to implement the annual reviews of domestic regulation that the USTR initially proposed. Last-minute nerves saw the USTR try to alter the agreement in favor of one grandfathering protectionist U.S. legislation, and AT&T's opposition to the final GATS agreement ensured that the United States withdrew most favored nation commitment to basic telecommunications. Even so, U.S. ratification had to rely on hasty passage by a "lame-duck" Congress.

The decision to negotiate on basic telecommunications was a U.S./E.U. decision. The ensuing negotiations were characterized by the conflicting interests of U.S. multinational users and U.S. operators, particularly AT&T. Withdrawal from the negotiations became a bargaining tool to satisfy domestic interests. As groups gained power within the United States, so they gained influence in the GATS negotiations—first the large multinational users, then AT&T, then the GMPCS and alternative carriers. The multinational users were initially successful in overriding AT&T's opposition to negotiations on basic telecommunications. But then AT&T was influential in gaining the U.S. exemption from the agreement's MFN clause and the FCC's introduction of the unilateral ECO bilateral reciprocity test. GMPCS operators later successfully pressured for a U.S. withdrawal until there was a "critical mass" of market-access agreements.

Following a European decision to exempt some audiovisual services, the United States took a last-minute MFN exemption on direct broadcasting by satellites in order to give itself bargaining positions in relation to Canada and Argentina. Overall, under congressional and industry pressure, the U.S. attitude to the WTO was about bilateral reciprocity presented in a multilateral package—seemingly little different from the original "nuclear option" of GATT of the 1940s. But the major difference was now that "bilateral reciprocity," as imposed through FCC market access decisions, was sector specific and therefore antithetical to WTO commitments. Even during the negotiations, the FCC espoused a "Western Union" model of protection at home and expansion abroad.

Finally, to replace its failed annex, the USTR introduced an FCC-inspired regulatory reference paper. Perhaps reflecting its final Japanese drafting, the paper became predominantly about principles, rather than detailed regulatory mechanisms. Aimed to ensure transparency and to reassure foreign investors, the regulatory reference paper entrenched domestic regulators' discretionary power to regulate the sector according to national priorities. Initially, U.S. delegation members from the private sector were very enthusiastic. There is also some doubt as to whether developing countries fully understood that the term "cost-orientation" (an E.U. term) in regard to interconnection rates in the reference paper related both to domestic interconnection rates and international "accounting rates." The clause was later to provide the United States with an opening to use against others' domestic regulatory policy.

Starting from a point where U.S. multinational corporations, particularly those in finance and data processing, had demanded end-to-end networks and the breaking down of national regulation by a system of "Empire rules," the result of the WTO strengthened the very states that the United States had set out to diminish. It did not give multinationals rights over sovereign governments as NAFTA did. And despite U.S. attempts to have the OECD agree to such rights, in 1996 the Multilateral Agreement on Investment failed.

Instead of loosening state regulation, the WTO made it stronger. Countries had only to comply with WTO regulations, not go beyond them. It gave them power against excessive World Bank demands. The beneficiaries from the WTO were state regulators. The WTO agreement delivered the "WTO model" with domestic liberalization firmly under the control of national regulators. It had the effect of enhancing the FCC's own domestic power against the states. But the WTO was not the end of the U.S. "Empire project." Two further phases remained.

The first involved U.S. unilateral attempts to bypass the ITU's "accounting rates" and to use the issue so as to introduce international "resale" worldwide. At the root of the U.S. annual settlements deficit was the FCC's decision in 1986 to make its international settlements policy applicable to telephony. The policy created incentives for U.S. companies to expand their market share of outgoing traffic so as to get a proportionate return of incoming traffic (much as ITT tried to do in the 1950s). Operators increased country-direct and call-back services, and the U.S. settlements deficit ballooned. The FCC then chose to fight the issue on the basis that settlement rates were above "costs."

Faced with opposition and lack of consensus within the ITU, the FCC/ USTR first introduced the issue into the WTO negotiations in 1992. Defied by Brazil and India in alliance with the European Union, the United States then reintroduced the issue of "cost-based" interconnection tariffs into the WTO's Negotiating Group on Basic Telecommunications. In a group dominated by the industrialized West, it succeeded in their inclusion in the regulatory reference paper. However, the provision on transparency of "accounting rates" was subsequently dropped. Whether there was deliberate intent to mislead developing countries is not known, but at least one of those agreeing to incorporate the regulatory reference paper into its commitments was unaware that "cost-oriented interconnection" included international interconnection (i.e., "accounting rates").

Once the WTO had been signed, the FCC declared the unilateral imposition of new, lower benchmark rates together with a schedule that demanded even the poorest countries should comply fully within five years. It then effectively vetoed the adoption of an alternative ITU-T recommendation that would have reduced the impact on the poorest. Since the FCC argued that the ITU rates would have impeded worldwide liberalization, its own must have been designed to force least developing countries into privatization. It also rejected a system based on local interconnection that would have affected AT&T's revenues.

Despite stating in 1997 that its intent was to bring down U.S. international collection charges, the FCC then acted to protect the revenue of AT&T and other operators by allowing domestic international tariffs to rise. Nor did it press for the end of state-to-state "accounting rates." It did, however, use the issue to license foreign operators for international "resale," thereby using interconnection as a weapon to extend its indirect control overseas, as it had done in the 1940s.

The U.S. complaint against Mexico under the WTO agreement was also primarily aimed at demonstrating that the agreement made international "resale" compulsory. But although the U.S. complaint delivered "cost-oriented accounting rates" and leased lines, it could not force Mexico into allowing "resale." In other words, after fifteen years of negotiations, the WTO did not deliver an "Empire rules" model of end-to-end networks without national regulation. However, FCC domestic power was enhanced by implementation of the WTO as it had been by ITU membership.

Finally, we document two further "Empire rules" attempts that have been successful. In the first—the creation of ICANN—the generalized U.S. government opposition to the ITU as an intergovernmental institution became evident with the 1996 refusal to allow it to become involved in the regulation of the domain names of the Internet. Following the State Department's public humiliation of the ITU's director general, the United States traded off support for the private corporation ICANN against ITU's hosting of the World Summit of the Information Society. ICANN became the front for the widely promoted fiction of a "self-regulated" Internet. Attempts to make ICANN more democratic always tinkered with the reality of a private corporation under U.S. Department of Commerce control.

As developing countries gained self-confidence, so, through the ITU, they demanded an intergovernmental organization. But these demands seemed to have withered when, in 2005, in the face of strongly worded direct U.S. opposition, the European Union withdrew support.

However, after ten years of maintaining a directly controlled, non-accountable private corporation with global regulatory power, the political intervention of the George W. Bush administration in that regulation brought renewed international pressure. As in the previous case of Comsat, the Europeans and Canadians headed the demand for reform. For other reasons, Congress was also highly critical of ICANN's performance. The combination of domestic and international pressure led in 2006 to an MOU between ICANN and the Department of Commerce that presaged three years of lighter control and an undertaking that the organization would become fully independent in 2009. Nevertheless, the U.S. government still retains control over the root server of the Internet (and therefore technical power), and in 2007 there is no evident model of internationalization that will follow ICANN's so-called "privatization." The lesson of Comsat was that the United States retained informal, structural power within the organization long after changes in its constitution signaled the end of formal U.S. power. Hence, it is too early to suggest that this "Empire rules" attempt has been preempted by international alliance.

A final successful attempt at an "Empire rules" model has come since the WTO in bilateral agreements. The WTO Basic Telecommunications Agreement was presented in 1997 to the U.S. public as the adoption by the rest of the world of the U.S. 1996 Telecommunications Act. The post-WTO period was characterized at first by attempts on the part of the FCC and USTR to impose that view on the rest of the world. In pursuit of that aim, the FCC used bilateral reciprocity to control foreign direct investment into the United States, and between 1999 and 2003 the USTR "named" thirty-four countries as offending under Section 1377 of the 1988 Trade Act on telecommunications issues. U.S. policy reverted to the "Western Union" model.

The Clinton administration turned to free-trade agreements with Jordan, Singapore, and Chile, and under the Bush administration the USTR used "fast track" authority under the Trade Act of 2002 to drive forward on bilateral, regional, and multilateral negotiations—in what became known as the strategy of "competitive liberalization." Based on the "Empire rules" model, these agreements brought a "top down" approach to limit the discretion available for national regulators to favor national goals. Like NAFTA, they give rights over sovereign governments to private investors. They were intended to provide building blocs to give strategic advantage to U.S. negotiators within the next round of the WTO. Increasingly however these agreements have brought domestic opposition from civil society groups questioning whether they benefit either the United States or the foreign country.

In the meantime, the U.S. model of regulation—of "enhanced" and "basic" services—on which the WTO agreement was based, has come up against competition. The Clinton administration favored the Internet, subsidized it, and classified it as other than "telecommunications" in order to escape regulation. Yet as telecommunications operators migrate from circuit switching to networks based on Internet standards, this regulatory model, as with data and voice previously, becomes politically rather than technically motivated. And, again, the U.S. position in the WTO negotiations has boomeranged.

The United States took advantage of the fact that the WTO agreement defined "telecommunications" according to how each government wished, so when others failed to define the Internet as "telecommunications," it could have no complaint. Yet that definition affected multinationals. In 2001, INTUG demanded that the Internet be defined as "basic telecommunications" worldwide. Further, from 1999 the European Union put forward proposals that would abolish the term "value added network" and bring the Internet under the category of "basic telecommunications" or "data transmission." Both definitions would demand major changes in U.S. domestic regulation. Such an impact was not what the United States intended the WTO to achieve. The opposite was the case.

Meanwhile, the concentration of domestic companies, involving even the end of AT&T, takes the U.S. domestic market back to the 1940s and circles the wagons against the developmental states of China, India, and Brazil, and against European Union expansion. It remains to be seen whether, by linking trade to defense, the "war on terror" can now deliver the U.S. "Empire project" in telecommunications.

To sum up, then, the story told here is of a fifty-year-long and still ongoing U.S. "Empire project" to ensure that its corporations—both operators and users—gain supremacy over sovereign governments. The "Empire project" has seen the United States use all forms of direct and indirect power and many unilateral, bilateral, and multilateral strategies to pursue its aims. The contentions of this book, written in 2007, is that the U.S. goal of enforcing the liberalization of worldwide national telecommunications markets so that U.S. companies might operate "future-proof" end-to-end control has only been partially successful. "Empire rules" for "enhanced services" were part of NAFTA, allowing the spread of multinationals' private networks from Canada to Mexico and have been part of recent bilateral agreements. But the WTO did not enforce similar rules on voice telecommunications or the Internet. The "Empire project" has been hampered in other arenas by the negotiating strategies of countervailing blocs and by the conflict between U.S. domestic agencies and their clients.

Still, state-based networks are the dominant force in international telecommunications. Although Intelsat has privatized and the ITU and World Bank are emasculated, the WTO agreement actually structured the international market away from the end-to-end model posited by the proponents of "globalization" as inevitable. It once more placed the sovereign state center stage. The European Union has also become a powerful countervailing regulatory bloc, while China and India's economic success provides the spearhead for revitalized developing-country demands. Yet, as demonstrated in 2005, the U.S. determination to retain absolute control over the Internet and its recent bilateral "Empire rules" agreements reveals that U.S. attempts at an "Empire project" in communications are still ongoing.

Notes

Introduction

1. Although the European Union was previously known as the European Economic Community and the European Communities, for simplicity I use European Union throughout the book.

2. Archival sources: NACP: U.S. National Archives at College Park, Maryland; WUA: Western Union Archives at National Museum of American History, Washington, D.C.; PRO: Public Record Office at Kew Gardens, U.K.; PO: Post Office/BT Archives Holborn Telephone Exchange, London; CW: Cable & Wireless Archives, Porthcurno, Cornwall, U.K.; Proquest: Proquest Archives available at http://pqasb .pqarchiver/nytimes. Exchange rates available at http://eh.net/hmit/exchangerates.

3. A "natural monopoly" is one where it is economic for one entity to provide a whole range of products or services.

4. I am indebted to Maria Michalis for the term "future proof."

5. Reed Hundt, then FCC Chairman. "Concerning WTO Agreement on Telecom Services." 18 February 1997, cited in Naftel and Spivak (2000, 107).

Chapter 1: Opening Up the British Empire

1. CCIF— International Telephony Consultative Committee; CCIT— International Telegraph Consultative Committee; CCIR— International Radio Consultative Committee. CCIF and CCIT were combined into CCITT (International Telegraph and Telephone Consultative Committee) in 1952.

2. International Telegraph Convention (St. Petersburg 1875), PRO/HO257/9.

3. *New York Times* 4 January 1940, WUA19/609.

4. FCC. 1939. *Report on the Telegraph Industry.* Submitted to Senate Committee on Interstate and Foreign Commerce, WUA11/16B.

5. *New York Times* 4 January 1940, WUA19/609.

6. *Wall Street Journal* 9 July 1941, WUA19/609.

7. Cable & Wireless Memorandum 1942, CW/B1/93.

8. Sir Edward Wilshaw to Treasury, 27 February 1942, PRO/DO35/1800.

9. Clement Attlee, Dominions Office, to Wilshaw, 15 September 1943, CW/1/473.

10. Wilshaw to Treasury, 15 October 1943, PRO35/1800.

11. Post Office Memorandum, 18 April 1944, PO33/5802.

12. "Shall U.S. External Communications be Unified?" *Fortune* 5 May 1944, reproduced in U.K. Embassy, Washington, D.C., Monthly Communications Bulletin, June 1944, PRO/FO 371/4261.

13. Commonwealth Communications Council, Memorandum on U.S. Telecommunications Policy, 4 May 1944, PRO/FO371/4261.

14. Commonwealth Communications Council, Memorandum on Commonwealth Network of Wireless Communications, 15 April 1944, PO/33/5830.

15. British High Commission, Egypt to Foreign Office, 28 October 1944, PO33/5802.

16. Wilshaw to Treasury, 27 April 1945, PRO/DO36/7748.

17. Commonwealth Telecommunications Board 1945, Report of Constitution Committee, PRO/T/162/809.

18. *New York Times* 8 March 1942, WUA/19/611.

19. FCC Docket 19660 (1980), "In the matter of the application for merger of the Western Union Telegraph Company and Postal Telegraph Inc. ITT World Communications Inc: for Revision of the Formula for the Distribution of Outbound International traffic 1964," NACP/173/550.

20. *New York Herald Tribune* 25 April 1942, WUA/19/609.

21. *New York Sun* 13 May 1942; *PM* 7 December 1942, WUA/19/609.

22. Cited in *Removal of Divestment Requirement of Section 222 of Communications Act of 1934. S.3646,* 97th Cong., 2d sess., 29 August 1962: 1, WUA9/357B.

23. Communications Act 1934 as amended 1943, sect. 222(e)(1).

24. *New York Times* 29 July 1943, Proquest.

25. Federal Communications Commission, Docket No. 6517 Western Union and Postal Telegraph Merger, vol. 21, "Separate Report of the Commission on Formulas for the Distribution of International Traffic," 1943, 9, NACP173/550.

26. See, for instance, FCC Docket 9292 "Lawfulness of Agreements between Western Union, Globe Wireless Ltd. and Tropical Radio Telegraph Co. concerning Exchange of International Telegraph Traffic," 1952, 58-59, NACP 173/550.

27. *New York Times* 21 December 1954, Proquest.

28. *Wall Street Journal* 30 September 1955, Proquest.

29. 1945 Report of Commonwealth Telecommunications Conference, appendix 10, PRO/CAB133/35.

30. *Telecommunications Reports* 21 November 1944, PRO/FO 371/50528.

31. Recalcitrant countries included: Argentina, Chile, Columbia, Uruguay, and the

Netherlands West Indies. Venezuela fell into line November 16, 1944. FCC Docket No. 6676 & 66046. Proposed Report, 3n2, PRO/FO371/50528.

32. Foreign Office Memorandum, 19 January 1945, PRO/FO371/50528.

33. FCC News Release, 29 December 1944, PRO/FO371/50528.

34. FCC, Proposed Report and Order Docket 6676 & 6046, 27 December 1944, PRO/FO371/50528.

35. Foreign Office Memorandum, "Dispute between F.C.C. and Cable and Wireless Ltd.," circa 1945, PRO/FO371/50528.

36. Ibid.

37. Letter Ray C. Wakefield, FCC Commissioner, to Wilshaw, 16 March 1945, PRO/FO371/50528.

38. Report of Discussion between United States and British Delegations, 15 March 1945, PRO/FO371/50528.

39. Foreign Office Memorandum, 7 May 1945, PRO/T/162/1026.

40. Telegram U.K. Embassy, Washington, D.C., to Foreign Office, 18 May 1945, PRO/FO371/50528.

41. Telegram U.K. Embassy, Washington, D.C., to Foreign Office, 4 July 1945, PRO/FO371/50528.

42. AP Managing Directors Association (22 October 1942, 46) cited in Renaud (1985, 13).

43. "Shall U.S. External Communications Be Unified?" *Fortune* May 1944, PRO/FO371/42460.

44. *Christian Science Monitor* 21 September 1944, Proquest.

45. *New York Times* 9 October 1944, Proquest.

46. *Washington Post* 1 October 1944, Proquest.

47. U.K. Embassy, Washington, D.C., to Foreign Office, 30 June 1944, PRO/FO/371/42460.

48. *New York Times* 18 March 1944, Proquest.

49. *Wall Street Journal* 28 March 1944, Proquest.

50. *Wall Street Journal* 20 October 1943; 19 October 1944, WUA/19/610.

51. *Washington Post* 6 May 1945, Proquest.

52. U.K. Embassy, Washington, D.C., to Foreign Office, 18 October 1944, PRO/Cab 21/1748.

53. Commonwealth Communications Council. Memorandum by Telephone and General Trust Ltd., 26 May 1944, PRO/FO 371/4260.

54. U.K. Embassy, Washington, D.C., to Foreign Office, PRO/Cab 21/1748.

55. *New York Times* 20 January 1943, WUA/19/610.

56. *New York Times* 10 March 1944, Proquest.

57. U.S. Senate, Interstate Commerce Committee, Report by Subcommittee: International cooperation in communication. S. Report 1907, 79th Congress, 2nd Session, Senate Miscellaneous Reports IV, cited in Feldman (1975, 88).

58. Executive Order No. 9831 Abolishing the Board of War Communications,

Federal Register 12 FR 1363, Feb 26, 1947. Available at http://www.uhuh.com/laws/donncoll/eo/1947/EO9831.TXT.

59. Bermuda Telecommunications Conference 1945. Minutes of the Opening Meeting, 22 November 1945, PO121/466.

60. U.S. companies present were: American Cable and Radio Corporation; Western Union Telegraph Company; RCA; RCA Communications Inc.; Radio Marine Corporation of America; Press Wireless Inc.; AT&T; Tropical Radio Telegraph Company. Commonwealth companies represented were: Cable & Wireless Ltd.; Canadian Marconi Co.; Canadian National Telegraphs; Canadian Pacific Telegraphs; Amalgamated Wireless (Australasia) Ltd., PO121/458.

61. Article IV, Section 16 (ii) of the Bermuda Agreement, *Washington Telecommunications Reports* 4 June 1946, PO33/5831.

62. "Report in the Matter of Radiotelegraph Circuits between the United States and British Commonwealth and Certain Other Foreign Points." 12 FCC 526 (1947) cited in Oslund (1977, 152).

63. *FCC v. RCA Communications Inc.* 346 US 86 (1953), cited in Oslund (1977, 153).

64. British Embassy, Washington, D.C., to Foreign Office, 29 January 1946, PO121/458.

65. Foreign Office to British Embassy, Washington, D.C., 27 April 1946, PO121/459.

66. Report by Hugh Townsend, GPO, on his visit to Washington, D.C., 3 June 1946, PO121/459.

67. Draft Letter to Brigadier J. J. Deedes, British Embassy, Washington, D.C., July 1946, PO121/459.

68. *Telecommunications Reports* 14 (2) 5 September 1947, PO121/459.

69. "Loss to Cable & Wireless Ltd on account of Bermuda Reductions June 1946 to May 1947 inclusive," n.d. [ca. 1948], PO 121/459.

70. *Telecommunications Reports* 22 March 1946, PO121/458.

71. Cable & Wireless owned two cables between Canada and Britain (previous state-owned Imperial cables).

72. "Supplementary Agreement to revise Article II of the Agreement Annexed to the Final Act of the Commonwealth-United States Telecommunications Meeting," London, 12 August 1949, PRO/FO371/99799.

73. London Talks on Telegraph Rates, 18 July 1952, PRO/ FO371/99799.

74. Cordell Hull to President Roosevelt, ca. September 1943, enclosing Report of Special Committee on Communications Peace Terms, PRO/FO115/3571.

75. Department of State, Memorandum on Conversations regarding Post War Planning, August 1945, PO33/5831.

76. Commonwealth Telecommunications Conference 1945, Committee on International Arrangements, 20 July 1945, PO33/5831.

77. Post Office Memorandum, 4 June 1946, PO33/5831.

78. Notes from Townsend for the use of Minister of State in private discussion, 4 June 1946, PO33/5831.

79. Post Office to Foreign Office, 9 October 1945, PO33/5831.

80. Post Office Memorandum, 4 June 1946, PO33/5831.

81. *Washington Telecommunications Reports* 31 May 1946, PO 33/5841.

82. Telegram from Permanent U.K. Representative to U.N. to Foreign Office, 25 February 1947, PO102/44.

83. Gerald C. Gross, then vice director of the International Telecommunications Union; first American ITU Secretary General 1960.

84. Post Office Memorandum, 21 February 1947, PO102/44.

85. Report of the U.K. Delegation, International Telecommunications Conferences, Atlantic City 1947, Annex A, International Radio Conference, 18, PO33/5910.

86. Francis Colt de Wolf (State Department) to J. C. Willever, Vice-President, Western Union, 18 February 1938, Administrative Records of U.S. Delegation 1938; J. C. Willever to W. R. Castle (State Department), 22 September 1932. Telegrams Sent and Received from the Dept. of State 1932, NACP43/190/1.

87. Townsend to Charles Dixon, 20 March 1948, PRO/ CO937/42/4.

88. Canada, Ecuador, Iraq, Mexico, Panama, Peru, Philippines, Saudi Arabia, United States, Uruguay, Venezuela.

89. Walter Radius (State Department) to Wayne Coy, Chairman FCC, 15 March 1948. FCC, Correspondence of Executive Officer, NACP173/6/20.

90. Proposal Relating to Chapter VII of the Telegraph Regulations, CITT, Paris 1949. FCC, Chief Engineer: Records Relating to Negotiation of International Agreements Treaties and Conferences, NACP173/550.

91. Report of Chairman of the International Telegraph Revision Committee, Geneva, 17 January to 4 February 1949. FCC, Chief Engineer: Records relating to negotiations of international agreements, treaties and conferences, NACP173/550.

92. Report: United States Delegation to the International Telegraph Regulations Revision Committee, Geneva, January 1949. FCC, Chief Engineer: Records relating to negotiations of international agreements, treaties and conferences, NACP173/550.

93. J. Laffey (Fr.) cited in Minutes, 11th Meeting, International Telegraph Regulations Revision Committee, Geneva, 31 January 1949. FCC, Chief Engineer: Records relating to negotiations of international agreements, treaties and conferences, NACP173/550.

94. Cited in Statement by the Portuguese Delegate. International Telegraph Regulations Revision Committee, Records relating to negotiations of international agreements, treaties and conferences, NACP173/550.

95. Report of the Chairman of the U.S. Delegation to the International Telegraph and Telephone Conference, Paris 1949: 21. FCC Chief Engineer: Records relating to negotiations of international agreements, treaties and conferences, NACP173/550.

96. French Telegraph Company FCC 79-776 released 5 December 1979, cited in "In the matter of International Record Carriers' Scope of Operations in the Continental

United States, including possible revisions to the formula prescribed under Section 222 of the Communications Act." FCC Docket 19660, released 27 February 1980: 6, NACP 173/550.

97. Letter FCC to International Record Carriers, 11 March 1949. FCC Chief Engineer: Records relating to negotiations of international agreements, treaties and conferences, NACP 173/550.

98. Public Conference, 16 January 1950, to consider reservations to the International Telegraph Regulations as revised at the International Telegraph & Telephone Conference, Paris, May to August 1949. FCC Chief Engineer: Records relating to negotiations of international agreements, treaties and conferences, NACP 173/550.

99. Report of the Telephone Committee (Committee 2) 30 May 1949, International Telegraph and Telephone Conference, Geneva, 1949, FCC Chief Engineer: Records relating to negotiations of international agreements, treaties, and conferences, NACP 173/550.

100. Public Papers of the Presidents, Dwight D. Eisenhower, 1960: 1035–40. http:// coursesa.matrix.msu.edu/~hst306/documents/indust.html.

101. ITT Draft Press Release, ca. 1954, PO33/6097A.

102. Post Office Minutes of Meeting, 9 September 1954, PO33/6097A.

103. Post Office Memorandum, ca. October 1954, para. 26(b), PO33/6097A.

104. Ibid., para 10.

105. Report of Informal Discussion on United States–United Kingdom Telegraph Services, Washington, D.C., May 1956, annex C., para 6, PO33/6097C.

106. Note of meeting with Canadian Overseas Telecommunications Corporation, 21/22 November 1956, PRO/DO/160/47.

107. See "The First Transatlantic Telephone Cable." http://www.sigtel.com/tel_hist_tat1.html.

108. Post Office. "Informal discussions with AT&T," 7 November 1956, PRO/ DO/160/47.

109. Post Office. Note of meeting 21/22 November 1956, PRO/DO/160/47.

110. COMPAC Pacific cable: Canada to Australia via Hawaii, Fiji, and New Zealand was completed in December 1963. SEACOM cable: Australia via New Guinea, Guam, Hong Kong to Singapore, with a microwave connection to Malaysia, came into service March 1967. Onward links to Sri Lanka, India, and East and South Africa were never constructed because satellites were more economic.

111. Allocation of Frequencies in the Bands Above 890 Mc. 27 FCC. 359 (1959) recon. 29 FCC. 825 (1960).

112. Booz, Allen, and Hamilton. 1962. *Organization and Management Survey of the Federal Communications Commission,* cited in "Legislation Note" *Harvard Law Review* 76: 390–91.

113. See John Crellin, "Commercial Cable." http://cial.org.uk/cable50.htm.

114. *New York Times* 26 February 1964; 18 March 1964. Proquest.

115. "In the Matter of the application for merger of the Western Union Telegraph

Company and Postal Telegraph Inc. (1964) Complaint of ITT World Communications Inc for Revision of the Formula for the Distribution of Outbound International Traffic," 7–8, FCC Docket 19660 (1980). NAMP 173/550.

116. FCC Chairmen: E. William Henry (D) 1963–66; Dean Burch (R) 1969–74; Richard E. Wiley (R) 1974–77; Charles D. Ferris (D) 1977–81.

117. *Wall Street Journal* 26 April 1946; 4 October 1946, WUA 19/610.

118. "Chile 1964: CIA Covert Support in Frei Election Detailed; Operational and Policy Records Released for First Time." The National Security Archive, 24 September 2004. http://www2.gwu.edu/-nsarchive/news/20040925/index/htm.; Church Report, "Covert Action in Chile 1963–73," 18 December 1975. U.S. Senate Staff Report of the Select Committee to Study Government Operations with Respect to Intelligence Activities. 9th Cong. 1st Sess. http: //foia.state.gove/Reports/Church.

Chapter 2: Satellites and U.S. Unilateral Regulation

1. John F. Kennedy. *Public Papers of the Presidents of the United States* 1961, 530, cited in Smith (1976, 79).

2. Report of the Ad Hoc Carrier Committee to the FCC and the Minority Statement of the Western Union Telegraph Company. 12 October 1961, reproduced in Musolf (1968, 29–32).

3. The opposition group included Senators Russell Long, Ralph Yarborough, Maurine Neuberger, Ernest Gruening, and Albert Gore.

4. See Booz, Allen, and Hamilton, *Organization and Management Survey of the Federal Communications Commission* March 1962, cited in *Harvard Law Review* 1962: 390–91.

5. Quoted in Press Release of G. Griffith Johnson, Asst. Sec. of State for Economic Affairs, before Subcommittee No. 4 of the Committee on Science and Aeronautics of the House of Representatives, 27 September 1962, PO/TCB2/182.

6. Post Office Memorandum, 14 August 1962, PO/TB2/182.

7. Post Office to J. E. Baldwin, Deputy Minister of Transport, Ottawa, 10 August 1962, PO/TB2/182.

8. Post Office Memorandum, "British & Canadian suggestions regarding tactics during the Washington discussions," December 1962, PO/TB2/182.

9. Ibid.

10. Department of State. *Summary of Activities of Department of State relating to the Communications Act of 1962.* September 20, 1963, cited in Galloway (1972, 93).

11. CEPT: an association of European PTOs established in 1959 as a nonministerial and nonpolitical body to improve the quality and range of public services. It met in plenary session with working parties and study groups. See Memo "Meeting of European PTT Ministers May 1969," PRO/HO255/1057.

12. Post Office Memorandum, December 1962, PO/TB2/182.

13. Outward Telegram to Commonwealth High Commissioners, 4 July 1963, PRO/DO160/58.

14. Foreign Office Telegram "European Conference on Satellite Communications," 26 July 1963, PRO/DO160/59.

15. John W. Finney. 1963. "Europeans Ask Managing Role in Space Communications Plan" *New York Times* 13 November, Proquest.

16. John W. Finney. 1964. "British May Join Satellite Corp." *New York Times* 26 January, Proquest

17. Postmaster General, Reginald Bevins, cited in *New York Times* 7 April 1964, Proquest.

18. "The 'space segment' refers to the tracking, control, command and related facilities and equipment required to support the operation of satellites." Note by Foreign and Commonwealth Office, February 1969, PO/TB2/54/3/26.

19. "Agreement Establishing Interim Arrangements for a Global Commercial Communications Satellite System," PRO/FO/99/1/752.

20. Gaunnady Stachevski, cited in U.K. Mission to U.N. to Foreign Office, 28 August 1967, PRO/FCO/55/671.

21. General Starbird. Evidence to House of Representatives, 1963, 6.

22. Evidence to House of Representatives, 1965, 101–2.

23. *New York Times* 19 July 1964. Proquest.

24. British Embassy, Washington, D.C., to Foreign Office. PRO/FCO/55/141.

25. Foreign Office and National Industrial Space Committee, 13 May 1968, PRO/Cab/164/119.

26. Foreign and Commonwealth Office Memorandum, 7 October 1965, FCO/371/183355.

27. Foreign Office Telegram to U.K. Embassy, Washington, D.C., 15 April 1965, PRO/FO371/183355.

28. Christopher Lydon. 1969. "Intelsat Proving Thorny Problem" *New York Times* 27 March, Proquest.

29. See, for instance, Post Office Report: Steering Committee on Satellite Communications, Interim Communications Satellite Committee. 42nd meeting, 13–20 August 1969; 28 August 1969, PO/TH/DA/679.

30. John Johnson, Vice President, Comsat, cited in Abram Chayes, "Speech to Stanford University Conference on Peaceful Uses of Outer Space." 16–18 August 1967, PRO/FCO/55/141.

31. President Johnson, "Message on Communications Policy to Congress," 14 August 1967, PRO/FCO/55/141.

32. British Embassy, Washington, D.C., to Foreign Office, 1 March 1967, PRO/FCO/55/141.

33. Post Office Memorandum, 23–24 November 1967; 27 November 1967, PRO/CAB/164/119.

34. Foreign Office Memorandum, 26 January 1968, PRO/CAB164/119.

35. Post Office Memorandum, 14 February 1968, PRO/CAB164/119.

36. See Thomas J. Hamilton, "Soviets to Establish Space Communications Network to Compete with Intelsat" *New York Times* 15 August 1968, Proquest.

37. Foreign and Commonwealth Office Memorandum, 11 June 1971, PO/T/3649/69.

38. Telegram, Foreign and Commonwealth Office to Rome Embassy, 8 January 1970, PRO/FCO/607.

39. Foreign and Commonwealth Office Memorandum, 19 June 1970, PRO/FCO 55/603.

40. Telegram, Foreign and Commonwealth Office to Washington, D.C., Embassy, 2 February 1970, PRO/FCO/55/604; U.S. Department of State, "Talking Points Concerning U.S. views with respect to the February 16 INTELSAT Conference," Washington, D.C., ca. January 1970, PRO/FCO/55/605; Telegram, Intelsat delegation to Foreign and Commonwealth Office, 4 June 1970, PRO/FCO/55/603.

41. Foreign and Commonwealth Office, "Report on the Intersessional Working Group, 25 November–18 December 1970," PO/T3649/69.

42. Cited in letter Foreign Office to Post Office, 11 August 1970, PRO/FCO/55/598.

43. Foreign and Commonwealth Office to Post Office, 21 December 1970, PO/T/3649/69.

44. Telegram, Intelsat delegation to Foreign and Commonwealth Office, 19 December 1970, PO/T/3649/69.

45. Telegram, British Embassy, Washington, D.C., to Foreign and Commonwealth Office, 8 June 1970, PO/T/364/9/69.

46. Foreign and Commonwealth Office Memorandum, 19 June 1970, PRO/FCO55/603.

47. Meeting between Post Office and State Department, 3 September 1970, PO/T/364/9/69.

48. Post Office to Ministry of Technology, 7 April 1970, PRO/FCO/55/594.

49. Executive Order No. 10530, 10 May 1954. Available at http://www.fcc.gov/ib/pd/pf/clla.html.

50. New York Times 16 July 1965, Proquest.

51. Merrill, Lynch, Pierce, Fenner & Smith, Inc. et al., Prospectus: 10,000,000 Shares, Comsat Satellite Corporation June 2 1964, 27, cited in Kinsley (1976, 65).

52. New York Times 5 March 1966, Proquest.

53. FCC Docket 68-212, 18 February 1968, cited in Borchardt (1970, 55).

54. Memorandum for the Hon. Dean Burch, Chairman, Federal Communications Commission, White House Press Release, January 23 1970, PRO/FCO/55/601.

55. Ibid.

56. New York Times 6 July 1971, Proquest.

57. 30 FCC. 2d 571, 1971, cited in comments of Western Union International Inc. to Third Notice of Inquiry. Docket 18875, NACP 173/550/36.

58. ITT v. FCC, 699F. 2nd 1219 (D.C.Cir.1983) cited in Geller (1984, 67).

59. Office of Telecommunications Policy, U.S. Executive Office of the President, "Recommendations on International Telecommunications Facilities Planning," 12–16 December 1975, cited in Pelton (1977, 109).

60. "Guidelines for TAT-6 Cable Utilization." FCC Docket 18875, 19 February 1976, cited in Pelton (1977, 110).

61. *Telecommunications Reports* 22 March 1976.

62. Some South American countries and Hawaii complained at the proliferation of cables. See Kinsley (1976, 77 and 109).

63. Stuart Eizenstat, president Carter's Chief Advisor on Domestic Affairs, cited in Hill (1979, 12).

64. Separate statement of Charles D. Ferris, Chairman, "In the matter of International Record," FCC Docket 19660, NACP 173/550.

65. Cited in U.S. Congress/House of Representatives 1981, 45.

66. Both the draft and final versions of Buckley's letter are reproduced in U.S. Congress/House of Representatives 1982, 137–42.

67. The five further applicants were RCA American Communications; International Satellite; PanAmerican Satellite; Cygnus Satellite, and Fininsat. *Financial Times* 21 January 1986.

68. See Rep. Timothy Wirth, *Congressional Record* H.3058, 9 May 1985.

69. FCC Docket No.87-67 "Policy for the Distribution of United States International Carrier Circuits among Available Facilities during the Post—1988 Period," Adopted 24 March 1988. Released 14 April 1988.

Chapter 3: International Market Structure and the ITU

1. CCITT, 1964–68. Study Group III, Contribution 22—E, 29 November 1966.

2. CCITT, 1964–68. Study Group III Contribution 4—E, 17 May 1966.

3. CCITT, 1964–68. Study Group III Contribution 21—E, 29 November 1966.

4. Recommendation D.6 (1984) CCITT Red Book Applicable after the 8th Plenary Assembly, Volume II.

5. Interviews, Tokyo, August 1985 and 1987; Washington, D.C., Spring 1988.

6. The Association of Data Processing Service Organizations (ADAPSO), founded 1961, became the Information Technology Association of America (ITAA) in 1991, a nonprofit trade association representing computer software and services companies.

7. Colombia claimed at WATTC that it was responsible for the creation of Article 9.

8. INTUG was given official observer status in the ITU in 1979.

9. The Lomé Convention is an international aid and trade agreement between African Caribbean and Pacific Countries and the European Union. See http://homepages .uel.ac.uk/myeo278s/ACP.htm. U.S. private-sector interests at WATTC told the author that they had been instructed to lobby certain African delegations, but had not done so because they did not know who the delegations were.

10. Northern Telecom, Nynex, City Corp., Electronic Data Systems, Comsat, IBM, AT&T, and the Association of Independent Telephone Companies, as well as smaller firms.

11. ITU WATTC, 1988, Document, 13-E, 2 August 1988.

12. ITU/CCITT COM III, Document R 1-E, 24–28 April 1989; Document R 2-E, 10–12 October 1989.

13. ITU/CCITT COM III, Document R 10-E, Annex 4, 23–25 May 1990.

14. ITU/CCITT COM III, Document R 10-E, Annex 3, 23–25 May 1990.

15. ITU/CCITT COM III, Document R 14-E, Draft Revision Recommendation D.1.

Chapter 4: Markets and Membership

1. ITU Plenipotentiary (1973) Document No.261/PC1973, cited in Kazuka (1989, 60).

2. Bowie, N. A. "Third World Countries: Positions and Achievements," presented at Conference on World Communications: Decision for the Eighties, Annenberg School of Communications, University of Pennsylvania, Philadelphia, May 1980 (p. 6), cited in Valentine (1981, 159).

3. By 1988, the Telecommunications Training Institute had provided technical training to more than fourteen hundred people from developing countries (Taylor 1989, 6).

4. Total financial contributions were 1,883,000 CHF in 1986 ($1,046,111); 2,360,000 CHF in 1987 ($1,583,892); 3,038,600 CHF in 1988 ($2,081,233).

5. ITU/BDT, "A Policy Perspective on the Role of the ITU in Development." World Telecommunication Development Conference, 21–29 March 1994, Doc. 3-E: 7–8.

6. The members of the Advisory Group were Paul Hansen (Chair), Rita Cruise O'Brien, Lynne M. Gallagher, Dale Hatfield, William H. Melody, Terrefe Ras-Work, Mahendra Pratap Shukla, Gabriel Tedros, and Björn Wellenius.

7. ITU/BDT, "A Policy Perspective on the Role of the ITU in Development." World Telecommunication Development Conference, 21–29 March 1994, Doc. 3-E: 12.

8. ITU Plenipotentiary Conference, Nairobi 1982, Document 485-E: 3, cited in Codding (1990, 141).

9. ITU Plenipotentiary Conference, Nice 1989, Resolution No. 55.

10. ITU Plenipotentiary Conference, Nice 1989, Document 507-E, 29 June 1989: R.2/12 cited in Codding (1990, 145). The committee included Algeria, Mali, Morocco, Senegal, Zimbabwe, Indonesia, China, India, and Saudi Arabia.

11. Note by the Secretary General. 1992. "World Bank." Additional Plenipotentiary Conference, Geneva, December. Document 16-E: 8.

12. In 1993, UNDP announced a decrease in contributions by 33 percent from 1994 to 1996. Telecommunications and the ITU would receive only a small percentage of UNDP funds.

13. ITU/BDT, "Proposals for the Work of the Conference: New Initiatives to Foster Telecommunication Development in the LDCs." World Telecommunication Development Conference, 21–29 March 1994, Doc 94/DT/14-E: 7.

14. ITU/BDT, "Proposals for the Work of the Conference: New Initiatives to Foster Telecommunication Development in the LDCs." World Telecommunication Development Conference, 21–29 March 1994, Doc 94/DT/14-E: 8–14.

15. U.S. Statement, "Ministerial Meeting on Global Communications," Kyoto, Japan, 22 September 1994: 99.

16. Africa One failed: after a number of sales the eventual owner was convicted of fraud in the WorldCom scandal.

17. Loral Qualcomm Satellite Services' Globalstar, TRW Inc's Odyssey, MCH's Ellipsat, and Constellation Communications's Aries.

18. International Telegraph and Telephone Consultative Committee (CCITT), 1968, *White Book Volume IIA*. IVth Plenary Assembly Mar del Plata, Annex 2 to Recommendation E250: 8 para 4, cited in Ergas et al. (1989, 8).

19. For Europe and the Mediterranean Basin, rates of remuneration were determined by the Tariff Group for Europe and the Mediterranean Basin (TEUREM). Accounting rate shares, determined on the basis of costs, were distance related. See OECD (1994, 9).

20. Treasury Memorandum, 3 December 1958, PRO/FO 35/7779.

21. Post Office Memorandum, 22 July 1970, PRO/T/319/2038.

22. Treasury Memorandum, 28 April 1970, PRO/T/319/2038.

23. ITU. 1990. *Follow-Up Study of the Costs of Providing and Operating International Telephone Service Between Industrialised and Developing Countries*. ITU, Geneva, cited in Aamoth (1992, 50–51).

24. FCC Report, *International Accounting Rates and the Balance of Payment Deficit in Telecommunications Services*, December 1988, cited in *FCC Week*, 17 April 1989.

25. Reproduced in OECD 1994, 106–8.

26. See for instance: letter Rep. Tom Bliley, Chair, U.S. House Commerce Committee to Esther Dyson, ICANN. 22 June 1999. http://www.icann.org/correspondence/Bliley-letter-22june99.htm.

27. Condoleeza Rice and Carlos M. Guiterrez. 2005. Letter to Jack Straw. 7 November. Reproduced in *The Register* 2 December 2005. http://www.theregister.co.uk.

Chapter 6: GATT/WTO and Telecommunications

1. Broadly defined, transborder data flows are movements across national boundaries of machine-readable data for processing, storage, or retrieval. See Sauvant (1983, 1).

2. The Trade Expansion Act of 1962 required the president to appoint a Special Representative for Trade Negotiations and established an interagency trade organization. Executive Order 12188 of 1980 renamed it the Office of the United States Trade Representative, expanded the agency, and assigned it overall responsibility for developing and coordinating implementation of U.S. trade policy. http://www.ustr.gov.

3. See OECD Declaration on Transborder Data Flows, 11 April 1985. http://www.oecd.org.

4. Harry L. Freeman, Senior Vice President, American Express. Statement in evidence to U.S. Congress. House of Representatives. 1983, 43–44 (emphasis added).

5. "Fast track" authority prevents Congress from amending trade agreements.

6. "Green room" refers to the color of the GATT Director General's room.

7. Group of Negotiations on Services. "Note on the meeting of 26–30 March 1990." MTN.GNS/32 Sec. 72. All WTO documents are available at: http://docsonline.wto .org.

8. Group of Negotiations on Services. "Note on the meeting of 26–30 March 1990." MTN.GNS/32 April 24 Sec. 59.

9. Draft Final Act embodying the results of the Uruguay Round of Multilateral Trade Negotiations Part 3. MTN.TNC/W/35.3 26 November.

10. Group of Negotiations on Services. Working Group on Audiovisual services. "Note on the meeting of 27–8 August 1990." MTN.GNS/AUD/1 Sec. 35.

11. Group of Negotiations on Services. "Note on the meeting of 10–25 July 1991." MTN.GNS/44 17 September.

12. For these demands, see lobbyist Harry L. Freeman, cited in *Transnational Data And Communications Report* 1991b, 14(6): 3–4.

13. Group of Negotiations on Services. 1991. Communication from the United States "United States Proposal on Telecommunications Services Liberalization and Most Favoured Nation (MGN)."

14. See NAFTA text. http://www-tech.mit.edu/Bulletins/Nafta/13.telecom.

15. The original members were the United States, Canada, Australia, New Zealand, Japan, South Korea, Malaysia, Singapore, Philippines, Thailand, Indonesia, and Vietnam.

16. Uruguay Round Agreement: Ministerial Decision. "Declaration on the Relationship of the WTO with the IMF." Marrakech. 15 April 1994.

17. Summary of GNS Informal Working Group Meetings September 20–24 on Basic Telecommunications Services. Prepared by Swedish delegation. Personal communication.

18. WTO. "Uruguay Round Decision on Negotiations on Basic Telecommunications." Adopted 15 April 1994. Marrakech, LT/UR/D-5/4.

19. Structural separation divides competitive from monopoly services when both are provided by a dominant telecommunications operator so as to prevent unfair cross-subsidization. Accounting separation is less severe but with the same intent. Number portability ensures consumers can take numbers with them when they migrate between networks.

20. WTO (1994) 27 July "Report of the meeting of 11 July 1994," TS/NGBT.

21. For an outline of this paper, see "The Regulator: U.S. Experience," *ITU News* 4/97: 39–43.

22. WTO (1995) 10 March "Report of Meeting of 27–28 February 1995," S/NGBT/5.

23. WTO (1995) 21 April "Communication from the European Community and

its Member States. GATS Coverage of Telecom Regulatory issues," S/NGBT/W6; 23 May "Report of the Meeting of 26 April 1995," S/NGBT/6.

24. WTO (1995) 31 July "Communication from the United States," S/NGBT/W/12/Add.3.

25. World Trade Organization (1995) 16 October, "Communication from the European Communities and their Member States," S/NGBT/W/12/Add.10.

26. WTO (1996) 26 February "Communication from the United States," S/NGBT/W/12/Add.3/Rev.1.

27. World Trade Organization (1999) Council for Trade in Services. Background note by the Secretariat. "International Regulatory Initiatives in Services," S/C/W/97.

28. Cuban Liberty and Democratic Solidarity Act of 1996 (Helms-Burton Law).

29. WTO (1997) Group on Basic Telecommunications. "Report of the Meeting of 14 February 1997," S/GBT/M/8.

30. WTO. "Group on Basic Telecommunications," report of the meeting of 15 February 1997, S/GBT/M/9.

31. Fourth Protocol to the General Agreement on Trade in Services. Reference Paper adopted by the Council for Trade in Services, 30 April 1996.

32. Directive 97/33/EC of the European Parliament and of the Council on Interconnection in Telecommunications with regard to ensuring universal service and interoperability through application of the principles of Open Network Provision (ONP) 1997 O. J. (L.199).

33. WTO. "Report of the Group on Basic Telecommunications." 15 February 1997, World Trade S/GBT/4.

34. WTO. 2004. "Mexico—Measures Affecting Telecommunications Services" Document: WT/DS204/R. para. 7.97.

35. The FCC adopted three benchmarks—15 cents (U.S.) per minute for upper income countries; 19 cents per minute for upper-middle and lower-middle income countries; and 23 cents per minute for lower-income countries.

36. Telstra of Australia, India's Videsh Sancar Nigam Ltd. (VSNL), Singapore Telecom, Kokusai Denshin Denwa Co. (KDD) of Japan, Hong Kong Telecommunications Ltd. (owned by Cable & Wireless and the Chinese government) together with a consortium of Philippines companies.

37. U.S. D.C. Circuit Court: *Cable & Wireless P.L.C. v. Federal Communications Commission* No. 97-1612 (D.C. Circuit) reported by American Society of International Law, "International Law in Brief" January 11–15, 1999. http://www.asil.org/ilib/ilib0201.htm.

38. WTO. 2004. "Mexico—Measures Affecting Telecommunications Services" Document: WT/DS204/R.

39. Ibid. para 7.170–71.

40. U.S. Department of Justice Press Release. "Justice Department Clears Worldcom/MCI Merger after MCI Agrees to Sell its Internet Business." 15 July 1998. http://www.usdoj.gov.atr/public/press_releases/1998/1829.htm.

Bibliography

Aamoth, Robert. 1992. "U.S. Accounting Rate Policies in the Asia-Pacific Region." *Intermedia* 20 (4–5): 49–51.

———. 1994. "Accounting Rate Policy Impact on Telecom Carriers in DCs." *Transnational Data and Communications Report* 17 (4): 26–34.

Aaronson, Susan Ariel. 1996. *Trade and the American Dream.* Lexington: University of Kentucky.

Acheson, Keith, John F. Chant, and Martin F. J. Prachowny. 1972. *Bretton Woods Revisited.* Toronto: Macmillan.

Ackman, Dan. 2002. "House Committees to Investigate Global Crossing." *Forbes,* 13 March. http://www.forbes.com/companies/2002/03/13/0313topnews.html.

Adonis, Andrew. 1994. "World Bank Sets Telecom Aid Rules." *Financial Times,* 3 March.

Allen, Peter, and Derek Nicholas. 1989. "Telecommunications and the Melbourne Meetings (WATTC-88 and CCITT) and International Telecommunications User View." *Conference on Telecommunications and the Melbourne Meetings.* London: IBC.

Ambrose, William W., Paul Hennemeyer, and Jean-Paul Chapon. 1990. *Privatizing Telecommunications Systems: Business Opportunities in Developing Countries.* International Finance Corporation, Discussion Paper 10. Washington, D.C.: World Bank.

APEC (Asia-Pacific Economic Cooperation). 1994. *APEC Economic Leaders' Declaration of Common Resolve.* Bogor, Indonesia. 15 November. http://www.apec.org/apec/leaders_ _declarations/1994.html.

Aronson, Jonathan D. 1997. "Telecom Agreement Tops Expectations." In *Unfinished Business: Telecommunications after the Uruguay Round,* ed. Gary Clyde Hufbauer and Erika Wade, 15–26. Washington, D.C.: Institute for International Economics.

Aronson, Jonathan D., and Peter F. Cowhey. 1988. *When Countries Talk: International Trade in Telecommunication Services.* Washington, D.C.: American Enterprise Institute.

AT&T (American Telephone & Telegraph). 1995. "AT&T Says 'Africa One' Would Provide Major Economic Boost." News release. 8 March. http://www.att.com/news/0395/950308.cha.html.

Auerbach, Karl. 2001. "My Senate Testimony on ICANN of February 14, 2001." http://www.cavebear.com/archive/cavebear.

Babai, Don. 1988. "The World Bank and the IMF: Rolling Back the State or Backing Its Role?" In *The Promise of Privatization: A Challenge for U.S. Policy,* ed. Raymond Vernon, 254–75. New York: Council of Foreign Relations.

Barlow, Maude, and Tony Clarke. 1998. *MAI: The Multilateral Agreement and the Threat to American Freedom.* New York: Stoddart.

Barnett, Michael, and Richard Duvall. 2005. "Power in International Politics." *International Organization* 59 (1): 39–75.

Barr, Michael B., and Kerry A. Walsh Skelly. 1987. "Telecommunications Trade Issues Ring Loudly In Washington." *Telematics* 4 (5): 5–9.

Bell, Daniel. 1973. *The Coming of Post-Industrial Society.* New York: Basic.

Bellchambers, William. 1994. "Kyoto Is an Opportunity to Revitalise an Old and Honoured Organisation." *Intermedia* 22 (4): 32–35.

Benzoni, Laurent, and Raymond Svider. 1994. "Departing from Monopoly: Asymmetries, Competition Dynamics and Regulatory Policy." In Noam and Pogorel, 237–58.

Berg, Elliot, and Mary M. Shirley. 1987. *Divestiture in Developing Countries.* World Bank Discussion Paper 11. Washington, D.C.: World Bank.

Berger, Mark T., and Mark Beeson. 1998. "Lineages of Liberalism and Miracles of Modernisation: The World Bank, the East Asian Trajectory and the International Development Debate." *Third World Quarterly* 19 (3): 487–504.

Besen, S. M., and J. Farrell. 1991. "The Role of the ITU in Standardisation: Pre-Eminence, Impotence or Rubber Stamp?" *Telecommunications Policy* 15 (4): 311–21.

Blackhurst, Richard. 1998. "The Capacity of the WTO to Fulfill its Mandate." In Krueger, *WTO as an International Organisation,* 31–58.

Blouin, Chantal. 2000. "The WTO Agreement on Basic Telecommunications: A Reevaluation." *Telecommunications Policy* 24 (2): 135–42.

Bodenheimer, Susanne. 1971. "Dependency and Imperialism: The Roots of Latin American Underdevelopment." *Politics and Society* 1 (3): 327–57.

Borchardt, Kurt. 1970. *Structure and Performance of the U.S. Communications Industry: Government Regulation and Company Planning.* Boston: Harvard University.

Bouton, Lawrence, and Mariusz A. Sumlinski. 1997. "Trends in Private Investment in Developing Countries: Statistics for 1970–95." IFC Discussion Paper 31. http://www.ifc.org.

Braga, Carlos A. Primo. 1997. "Liberalizing Telecommunications and the Role of the World Trade Organization." *Public Policy for the Private Sector* 120 (1 July). http://rru.worldbank.org/PublicPolicyJournal/Summary.aspx?id=120.

Brandt Report. 1980. *North-South: A Programme for Survival; Report of the Independent Commission on International Development Issues.* Sydney: Pan.

Broache, Anne. 2006. "U.S. Government Renews Domain-Name Contract." *CNET News,* 17 August. http://news.zdnet.co.uk/communications/0,1000000085,39281139,00.htm.

Broadman, Harry G., and Carol Balassa. 1993. "Liberalizing International Trade in Telecommunications Services." *Columbia Journal of World Business* 4 (Winter): 30–38.

Brock, William F. III. [U.S. Trade Representative.] 1982. "Discusses the GATT Ministerial," 22 November; "An Opening Statement to GATT Ministers," 26 November. News release [combined]. U.S. Information Service.

Brockbank, R. A. 1958. "Overseas Telephone Cables since the War." *New Scientist* 3 (20 November): 1320–23.

Brown, Bartram S. 1992. *The United States and the Politicization of the World Bank.* London: Kegan Paul.

Browne, Robert S., and Robert J. Cummings. 1984. *The Lagos Plan of Action vs. the Berg Report.* Lawrenceville, Va.: Brunswick.

Brummer, Alex. 1990. "World Bank Improves Transfers to Developing Nations." *Guardian,* 17 September.

Business Round Table. 1985. *International Information Flow: A Plan for Action.* New York: Business Round Table.

Butler, R. E. 1987. "Looking to the Future: The ITU in the Year 2000." *The Legal Symposium: Harmonization of Global Telecommunications Systems.* 5th World Telecommunications Forum 87, 21–23 October. Geneva: ITU.

———. 1989a. "ITU Leadership in Achieving Global Interconnectivity." Speech to *Telecommunications and the Melbourne Meetings.* 31 March. London: IBC.

———. 1989b. "In Pursuit of Excellence: A Critical Choice." *Telecommunications Journal* 54 (May): 5–6.

Cable & Wireless plc. 1989. *Report and Accounts.* London: Cable & Wireless.

Carsberg, Bryan. 1985. "OFTEL—The Challenge of the First Five Years." *Information Technology and Public Policy* 4 (1) (September): 1–11.

Caufield, Catherine. 1996. *Masters of Illusion: The World Bank and the Poverty of Nations.* New York: Holt.

CCIF (International Telegraph Consultative Committee). 1954. *XVII Plenary Assembly Volume VI: 4–12 October 1954.* Geneva: ITU.

———. 1957. *XVIII Plenary Assembly Volume I bis of the Green Book: 3–14 December 1956.* Geneva: ITU.

CCITT (International Telegraph and Telephone Consultative Committee). 1964–68. *3rd Plenary Assembly: General Tariff and Lease of Circuits; Questions Entrusted to Study Group III for the Period 1964–68.* Geneva: ITU.

———. 1965. *Blue Book Applicable after Third Plenary Assembly Volume II: Series D; Recommendations (Study Group III)*. Geneva: ITU.

———. 1984. *Eighth Plenary Assembly, Malaga, Torrimolinos*. Geneva: ITU.

———. 1987. *Circular Letter 62*. 25 May. Geneva: ITU.

Chayes, Abram. 1971. "Unilateralism in United States Satellite Communications Policy." In McWhinney, 42–50.

Chomsky, Noam. 1991. *Deterring Democracy*. New York: Holt.

Clinton, William J. 1992. "The Economy." Speech. Wharton School of Business, University of Pennsylvania, Philadelphia, 16 April. http://www.ibiblio.org/nii/econ-posit.html#info.

Codding, G. A. 1972. *The International Telecommunication Union: An Experiment in International Cooperation*. New York: Arno.

———. 1984. "Politicization of the ITU: Nairobi and After." In *Policy Research in Telecommunications*, ed. Vincent Mosco, 435–47. Norwood, N.J.: Ablex.

———. 1989. "Financing Development Assistance in the ITU." *Telecommunications Policy* 13 (1): 13–23.

———. 1990. "The Nice Plenipotentiary Conference." *Telecommunications Policy* 14 (2): 139–49.

———. 1991. "Evolution of the ITU." *Telecommunications Policy* 15 (4): 271–85.

———. 1993. "After the December Plenipot: The New Look ITU." *Intermedia* 21 (2): 23–24.

Codding G. A., and Antony M. Rutkowski. 1982. *The International Telecommunication Union in a Changing World*. Dedham, Mass.: Artech.

Cohen, Edward S. 2001. *The Politics of Globalization in the United States*. Washington, D.C.: Georgetown University Press.

Colino, R. R. 1986. "An Inside Look at the 'New' Intelsat." *Telephony* 21 (12): 50–52.

Commission of the European Communities. 1987. *Towards a European Economy: Green Paper on the Development of the Common Market for Telecommunications Services and Equipment*. Brussels: EEC.

———. 1989. *Report on United States Barriers to Trade and Investment 1989*. Brussels: European Commission.

———. 1992. *Report on United States Barriers to Trade and Investment 1992*. Brussels: European Commission.

Commonwealth Telecommunications Bureau. n.d. [circa 1994]. Mimeo. *Commonwealth Collaboration in International Telecommunications*. London: Commonwealth Telecommunications Bureau.

Communications Week International. 1996. "FCC's Harris Wields Big Stick of Deregulation." 18 March: 1, 32–33.

Computergram International. 1998. "ITU Plays Down Its Role in Internet Governance." 9 November. http://www.findarticles.com/p/articles/mi_moCGN/is_3534/ai_53196096.

Cooper, Kent. 1942. *Barriers Down*. New York: Farrar.

Cowhey, Peter. 1987. "Trade in International Telecommunications and Information Services: An Agenda for the GATT Negotiations." In *The GATT Round*, I: 1–5. Paris: Prométhee.

———. 1989. "Telecommunications and Foreign Economic Policy." In *New Directions in Telecommunications Policy*, vol. 2, ed. Paula R. Newberg, 187–230. Durham, N.C.: Duke University Press.

———. 1990. "The International Telecommunications Regime: The Political Root of Regimes for High Technology." *International Organization* 45 (2): 169–99.

———. 1998. "FCC Benchmarks and the Reform of the International Telecommunications Market." *Telecommunications Policy* 22 (11): 899–911.

Cowhey, Peter, and J. A. Aronson. 1991. "The ITU in Transition." *Telecommunications Policy* 15 (4): 298–310.

Cukier, Kenneth. 1997. "OECD Urges Governmental Action over Domain Names." *Communications Week International*, 7 April: 1, 26.

Cullen, B. C. 1983. "The Users Role at the CCITT: An INTUG Perspective." *Telecommunications Journal* 50 (4): 260–63.

———. 1987. "Regulation and the User." *Telecommunications Journal* 54 (3): 180–83.

Curzon, Gerald, and Victoria Curzon. 1969. "GATT: Trader's Club." In *International Organisation: World Politics*, ed. Robert W. Cox, 298–333. London: Macmillan.

Dale, Reginald, and William Chislett. 1981. "Reagan Sets Out U.S. Stance at Summit." *Financial Times*, 22 October.

Dawkins, Will, and Hugo Dixon. 1989. "The Apron Strings Are Being Untied." *Financial Times*, 7 December.

Dawnay, Ivo. 1986. "Brazil Takes Softer Line Over Issue of Services in GATT Talks." *Financial Times*, 19 June.

Dixon, Hugo. 1990a. "Crossed Lines on a Tariff for International Calls." *Financial Times*, 23 April.

———. 1990b. "Phone Cartel Looking Vulnerable." *Financial Times*, 2 July.

———. 1990c. "BT and Mercury Deals Could Cut Transatlantic Call Charges." *Financial Times*, 13 October.

———. 1991a. "Cartel Called to Account." *Financial Times*, 29 January.

———. 1991b. "U.S. Acts to Halve Cost of International Phone Calls." *Financial Times*, 10 May.

Doran, Janis. 1989. *Middle Powers and Technical Multilateralism in the International System*. Ottawa: North-South Institute.

Downs, Anthony. 1957. *An Economic Theory of Democracy*. New York: Harper.

Drake, William J. 1988. "WATTC-88 Restructuring the International Telecommunications Regulations." *Telecommunications Policy* 12 (3): 217–33.

———. 1994. "Asymmetric Deregulation and the Transformation of the International Telecommunications Regime." In Noam and Gerald Pogorel, 137–205.

Drake, W. J., and E. M. Noam. 1997. "The WTO Deal on Basic Telecommunications." *Telecommunications Policy* 21 (9/10): 799–818.

Dryden, Steven J. 1992. "The Quiet Renewal of the Japan Chip Pact." http://aliciapatterson.org/APF1501/Dryden/Dryden.html.

Dullforce, William. 1985. "Gatt Poised to Clear Way for Trade Talks Next Year." *Financial Times,* 28 November.

Dunkel, Arthur. 1992. "Telecom Services and the Uruguay Round." *Transnational Data and Communication Report* 15 (1): 1/–19.

Dunkley, Graham. 1997. *The Free Trade Adventure: the WTO, the Uruguay Round and Globalism—A Critique.* New York: Zed.

Dunne, Nancy. 1985. "Germans Buy into Intelsat Network for Domestic Use." *Financial Times,* 13 December.

———. 1986. "Intelsat Keeps Competitors in the Cold." *Financial Times,* 21 January.

———. 1989a. "Hills Prevents Breakdown of Talks With Japan." *Financial Times,* 28 June.

———. 1989b. "Hills Offers a Liberal Line on Trade Policy." *Financial Times,* 16 August.

———. 1996. "Congress Warns Clinton over Telecoms." *Financial Times,* 29 April.

Eckes, Alfred E. Jr. 1999. "U.S. Trade History." In *U.S. Trade Policy: History, Theory and the WTO,* ed. William A. Lovett, Alfred E. Eckes Jr., and Richard L. Brinkman. New York: Sharpe.

Edwardson, Mickie. 1999. "James Lawrence Fly, the FBI, and Wiretapping—The Chairman of the Federal Communications Commission, 1939–44; Director of American Civil Liberties Union; Federal Bureau of Investigation." http://www.findarticles.com/p/articles/mi_m2082/is_2_61/ai_54469158/print.

Eger, John. 1981. "Law and Policy in International Business." In *Telecommunications in the U.S.A.,* ed. L. Lewin, 369–414. Dedham, Mass.: Artech.

Egger, Michel, and Jean-Louis Fullsack. 2003. "ITU, Swisscom and Development Cooperation: Neoliberalism versus Solidarity." *Developmentdotcom.* http://www.unige.ch/iued/wsis/DEVDOT/00853.HTM.

Ergas, Henry. 1994. "Look Behind the Telco Rhetoric." *Communications Week International,* 13 June: 14.

Ergas, Henry, and Paul Paterson, with Philip Geissler. 1989. "International Telecommunications Accounting Arrangements: An Unsustainable Inheritance?" Colloquium on the Development of International Telecommunications, Villefranche-sur-Mer, France, June 1–3.

Ergas, Henry, and Gerald Pogorel. 1994. "Multilateral Cooperation in International Telecommunications: Sources and Prospects." In Noam and Pogorel, 17–32.

Eward, Ronald. 1985. *The Deregulation of International Communications.* Dedham, Mass.: Artech.

Fatoumata, Jawara, and Aileen Kwa. 2003. *Behind the Scenes at the WTO: The Real World of International Trade Negotiations; The Lessons from Cancun.* New York: Zed.

FCC (Federal Communications Commission)/U.S. State Dept. 1943. *External Relations of the U.S.A.* 35–40. Washington, D.C.: GPO.

———. 1959. *External Relations of the USA*. 106–27. Washington, D.C.: GPO.

———. 1960. *External Relations of the USA*. 99–103. Washington, D.C.: GPO.

———. 1992. *Second Report and Order* FCC Rcd 8040. Washington, D.C.: FCC.

———. 1997. *Report and Order in the Matter of International Settlement Rates*. IB Docket No. 96–261 (benchmark order), 7 August.

Feketekuty, Geza. 1987. "Trade Policy Objectives in Telecommunications." In *World Telecommunication Forum: Part 3, Legal Symposium*," ed. ITU, 189–93. Geneva: ITU.

———. 1988. "Telecommunications Policy Reform and International Trade." *Financial Times World Conference*, 1 December: 1–13.

———. 1989. "Telecom Annex to Framework Agreement on Trade in Services," 11th IDATE Conference, Montpelier, France. *Transnational Data and Communication Report* 12 (10): 6.

Feldman, Mildred L. 1975. *The Role of the United States in the International Telecommunication Union*. N.p.: Feldman.

Fink, Carsten, Aaditya Mattoo, and Randeep Rathindran. 2001. "Liberalizing Basic Telecommunications: The Asian experience." World Bank Policy Research Working Paper. Washington, D.C.: World Bank.

Finnie, Graham. 1988. "The Changing Face of the ITU." *Telecommunications* 22 (10): 49–57.

———. 1995. "Rhetoric Meets Reality as Europe Confronts the GII." *Communications Week International*, 6 March: 1, 44.

Finnie, Graham, and Jennifer L. Schenker. 1995. "Global Info-Highways Idle." *Communications Week International*, 20 March: 1.

Fleming, Stewart. 1985. "U.S. to Examine 'Unfair' Foreign Trade Controls." *Financial Times*, 9 September.

———. 1986a. "World Bank on Brink of Taking in More Funds than Go in Aid." *Financial Times*, 23 June.

———. 1986b. "Wanted: A Strong Leader for a New Mission." *Financial Times*, 11 March.

———. 1986c. "Conable Changes Management at World Bank." *Financial Times*, 4 December.

Foley, Theresa. 1996a. "Latin America Faces U.S. Satellite Threat." *Communications Week International*, 15 July: 15.

———. 1996b. "A Sea of Activity on World's Optic Highways." *Communications Week International* 18 March: 17.

———. 1996c. "Satellite Numbers Set to Rocket." *Communications Week International*, 24 June: 8.

Frank, André Gunder. 1966. "The Development of Underdevelopment." *Monthly Review* (September): 17–30.

Fredebeul-Krein, Markus, and Andreus Freytag. 1999. "The Case for a More Binding WTO Agreement on Regulatory Principles in Telecommunications Markets." *Telecommunications Policy* 23: 625–44.

Freed, Ken. n.d. "From gTLD-MOU to ICANN." *Media Visions Journal.* http://www.media-visions.com/icann-gtld.htm.

Freedman, Des. 2006. "Media Policymaking in the Free Trade Era: The Impact of the GATS Negotiations on Audiovisual Industries." In *Trading Culture: Global Traffic and Local Cultures in Film and Television,* ed. Sylvia Harvey, 13–20. Eastleigh, U.K.: Libbey.

Friedman, Alan. 1996. "Chance of WTO Technology Deal Is Looking Good Again." *International Herald Tribune,* 10 December. http://www.iht.com/articles/1996/12/10/wto.t_7.php.

Galal, Ahmed, Leroy Jones, Pakaj Tandon, and Ingo Vogelsang. 1992a. *Welfare Consequences of Selling Public Enterprises: Questions and Approaches to Answers.* World Bank Conference, June 11–12. Washington, D.C.: World Bank.

———. 1992b. *Welfare Consequences of Selling Public Enterprises: Synthesis of Cases and Policy Summary, Draft.* 23 June. Washington, D.C.: World Bank.

Galbraith J. K. 1994. *The World Economy since the Wars: A Personal View.* London: Mandarin.

Galloway, Jonathan F. 1972. *The Politics and Technology of Satellite Communications.* Lexington, Mass.: Heath.

Galtung, Johan. 1971. "A Structural Theory of Imperialism." *Journal of Peace Research* 13 (2): 81–94.

Gardner, Richard N. 1956. *Sterling-Dollar Diplomacy.* Oxford: Clarendon.

———. 1972. "The Political Setting." In Acheson et al., 20–33.

Geller, Henry. 1984. "Chapter 4." In Sterling, 64–71.

Gilbert, Christopher L., Raul Hopkins, Andrew Powell, and Amlan Roy. 1997. "The World Bank: Its Functions and Future." ESRC Global Economic Institutions Discussion Paper. London: ESRC.

Goldschmidt, Douglas. 1984. "Financing Telecommunications for Rural Development." *Telecommunications Policy* 8 (3): 181–203.

Goldstein, Irving. 1992. "Intelsat's CEO Underlines Strategy." *Telecommunications* 26 (8): 12.

Gowers, Andrew. 1985. "Sacred Cows and the Gatt." *Financial Times,* 15 October.

Graham, George. 1993a. "Developing a More Worldly Bank." *Financial Times,* 2 July.

———. 1993b. "World Bank Approves Greater Transparency." *Financial Times,* 28/29 August.

Grieco, Joseph M. 1984. *Between Dependency and Autonomy: India's Experience with the International Computer Industry.* Berkeley: University of California Press.

Grub, Philip. 1984. "Can a Trade War Be Avoided?" In Sterling, 139–47.

Guest, Iain. 1984. "Gatt Compromise on Trade in Services." *Guardian,* 1 December.

Gwin, Catherine. 1993. "Wanted: World Bank Leadership." *Economic Insights* 4 (5): 9–12.

———. 1997. "U.S. Relations with the World Bank, 1945–1992." In *The World Bank: Its First Half Century,* vol. 2, eds. Devish Kapur, John P. Lewis, and Richard Webb, 195–274. Washington D.C.: Brookings.

Haas, P. M. 1992. "Introduction: Epistemic Communities and International Policy Coordination." *International Organization* 46 (1): 1–35.

Hamelink, Cees. 1994. *The Politics of World Communication.* London: Sage.

Hansell, Kathleen J., and Fran P. Putney. 1989. "Choosing Transatlantic Digital Services." *Telecommunications* 23 (6): 43–44.

Hardy, Michael. 1987. "Telecommunications in the European Economic Community." *World Telecommunication Forum, Part 3, Legal Symposium,* 195–97. Geneva: ITU.

Harmon, Amy. 1998. "U.S. Expected to Support Shift in Administration of the Internet." *New York Times,* 19 October.

Harrod, Roy. 1972. "Problems Perceived in the International Financial System." In Acheson et al., 5–19.

Hart, Kenneth. 1996a. "Internet Address Sell-Off Plan." *Communications Week International,* 15 July: 7.

———. 1996b. "Domains to Help Avoid Names Clash." *Communications Week International,* 15 July: 3.

Harvard Law Review. 1962. "Legislation Note: The Communications Satellite Act of 1962." 76: 390–91.

Hellman, Donald C. 1997. "America, APEC, and the Road Not Taken: International Leadership in the Post–Cold War Interregnum in the Asia Pacific." In *From APEC to Xanadu: Creating a Viable Community in the Post–Cold War Pacific,* eds. Donald C. Hellman and Kenneth B. Pyle, 70–97. New York: Sharpe.

Heyworth, Laurence. 1995. "Telcos Face Financial Drain Post-Privatization." *Communications Week International,* 6 March: 10.

Hill, Alice, and Manuel Angel Abdala. 1993. *Regulation, Institutions and Commitment: Privatization and Regulation in the Argentine Telecommunications Sector.* Policy Research Working Paper 1216. Washington, D.C.: World Bank.

Hill, Arthur. 1979. "Carter Administration Unveils Communications Policy." *Satellite Communications* 3 (11): 12.

Hills, Jill. 1984. *Information Technology and Industrial Policy.* Beckenham, U.K.: Helm.

———. 1986. *Deregulating Telecoms: Competition and Control in the United States, Japan and Britain.* London: Pinter.

———. 1989. "The Internationalisation of Domestic Law in Telecommunications." In *European Telecommunications Policy Research,* ed. N. Garnham with A. Askoy, 46–58. Netherlands: IOS.

———. 1991. "Regulation, Administration and Politics." In *Deregulating Regulators,* ed. Jean-Paul Chamoux, 99–107. Amsterdam: IOS.

———. 1993. "Economics as Ideology: The World Bank and Privatisation." Paper presented at the Center for International Business and Research, Michigan State University, May.

———. 1998a. "Liberalisation, Regulation and Developing Countries." *Gazette* 60 (6): 459–76.

———. 1998b. "International Accounting Rates: The Process of Demise of an International Regime." *Communications and Strategies* 30 (2nd Q): 241–69.

———. 2002. *The Struggle for Control of Global Communication: The Formative Century.* Urbana: University of Illinois Press.

———. 2007. "Searching for Universal Access: The Public Interest, the FCC, and the Regulation of International Telecommunications." *info* 9 (2/3): 83–96.

Hills, Jill, with Stelios Papathanassopoulos. 1991. *The Democracy Gap: The Politics of Information and Communication Technologies in the United States and Europe.* Westport, Conn.: Greenwood.

Hinchman, Walter R. 1971. "The Technological Environment for International Communications Law." In McWhinney, 21–41.

Hirobe, Kazuya. 1987. "Harmonization between National Telecommunications Law and International Telecommunications Law." *Legal Symposium: Harmonization of Global Telecommunications Systems,* 21 October, 15–17. Geneva: ITU.

Hirsch, Alan. 1998. "Comment on the Paper by J. Michael Finger and L. Alan Winters." In Krueger, *WTO as an International Organisation,* 392–97.

Hirst, Paul Q., and Grahame Thompson. 1999. *Globalization in Question.* 2nd ed. Cambridge: Polity.

HMSO (Her Majesty's Stationery Office). 1964. *Satellite Communications.* Cmnd. 2436. London: HMSO.

———. 1973. *Commonwealth Telecommunications Organisation Financial Agreement 1973.* Cmnd. 5319. London: HMSO.

Hoffman, David. 1990. "Study Urges Soviet Shift to Market Economy: International Agencies in Exhaustive Report, Say Poland-Style 'Shock Therapy' Is Needed." *Washington Post,* 18 December.

Home Office Radio Regulatory Department. 1980. *Report on the World Administrative Radio Conference 1979.* London: HMSO.

Horten, Monica. 1992. "Eastern Europe: A Magnet for Suppliers." *Financial Times,* 15 October.

Hosein, Gus. 2003. "The Sources of Laws: Policy Dynamics in a Digital and Terrorized World." *The Information Society* 20 (3): 187–99.

Hudson, Heather, et al. 1979. *The Role of Telecommunications in Socio-economic Development: A Review of the Literature with Guidelines for Further Investigation.* Washington, D.C.: Keewatin.

Hudson, Heather. 1983. "The Role of Telecommunications in Development: A Synthesis of Current Research." In *Proceedings from the Tenth Annual Telecommunications Policy Research Conference,* eds. O. H. Gandy Jr., P. Espinos, and J. Ordover, 291–308. Norwood, N.J.: Ablex.

———. 1984. *When Telephones Reach the Village: The Role of Telecommunications in Rural Development.* Norwood, N.J.: Ablex.

Hundt, R. 1997. "From Buenos Aires to Geneva and Beyond." Speech. World Affairs Council, Philadelphia, 22 October.

———. 2000. *You Say You Want a Revolution: A Story of Information Age Politics.* New Haven, Conn.: Yale University Press.

ICANN (Internet Corporation for Assigned Names and Numbers). 1998a. "Icann

Bylaws for Internet Corporation for Assigned Names and Numbers." http://www
.icann.org/general/archive-bylaws/bylaws-06nov98.htm#VII.

———. 1998b. "Articles of Incorporation of Internet Corporation for Assigned Names
and Numbers (as revised November 21, 1998). http://www.icann.org/general/
articles.htm.

ICANNWatch. n.d. "ICANN for Beginners." http://www.icannwatch.org/
icann4beginners.shtml.

IFC (International Finance Corporation). 2004. "Investment in Infrastructure Projects
with Private Participation in Developing Countries 1995–2004." http:www-wds
.worldbank.org/servlet/WDSContentServer/WDSP/IB/2005/1 1/08/00.

infoDev. 2000. "World Regulatory Colloquium for the Networked Economy. Guidelines for Applications." http://www.infodev.org.

Intelsat. 1986. *Bridging the Gap II*. Washington, D.C.: Intelsat.

———. 1991. *Annual Report 1990–91*. Washington, D.C.: Intelsat.

———. 1996. *The Agreement and Operating Agreement: Signed 20 August 1971; Entered
into Force 12 February 1973*. Washington, D.C.: Intelsat.

INTUG (International Telecommunications User Group). 2001. "INTUG Position
Paper: WTO General Agreement on Trade in Services." June. http://www.intug
.net/views/wto_gats.html.

Irwin, Michael H. K. 1990. "Banking on Poverty: An Insider's Look at the World
Bank." *Cato Foreign Policy Briefing No. 3.* http://www.cato.org/pubs/fpbriefs/
fpb-003.html.

Ito, Yukio. 1993. "International Simple Resale Carrier Wars." *Transnational Data and
Communications Report* 16 (2): 12–14.

ITU (International Telecommunication Union). 1947. *International Telecommunication Convention* (Atlantic City 1947). Geneva: ITU.

———. 1949. *Telephone Regulations (Paris Revision 1949) Annexed to the International
Telecommunication Convention* (Atlantic City 1947). Geneva: ITU.

———. 1953. *International Telecommunication Convention* (Buenos Aires 1952). Geneva: ITU.

———. 1959. *Minutes of Ordinary Administrative Telegraph and Telephone Conference*. Geneva: ITU.

———. 1973. *Final Acts of the World Administrative Telegraph and Telephone Conference (Geneva 1973): Telegraph Regulation; Telephone Regulations*. Geneva: ITU.

———. 1985. *The Missing Link: Report of the Independent Commission for Worldwide
Telecommunications Development*. Geneva: ITU.

———. 1988. *Final Acts of the World Administrative Telegraph and Telephone Conference, WATTC-88*. Geneva: ITU.

———. 1989. *The Changing Telecommunication Environment: Policy Considerations
for the Members of the ITU*. Geneva: ITU.

———. 1991. *Tomorrow's ITU: The Challenge of Change*. Report of High-Level Committee. Geneva: ITU.

———. 1996. *The African Green Paper*. Geneva: ITU.

———. 1997a. "Meeting on Changes to Internet Domain Name Structure Begins. 30 April. http://www.itu.int/newsarchive/projects/dns-meet/DNS-PressNote.html.

———. 1997b. *World Telecommunications Development Report*. Geneva: ITU.

———. 1999a. *Trends in Telecommunications Reform*. Geneva: ITU.

———. 1999b. "ITU Becomes a Founding Member of the ICANN Protocal Supporting Organization." News release. July. http://itu.int/newarchive/press_releases/1999/11.html.

———. 2002. "Report on ITU Activities Related to Resolution 102: Management of Internet Names and Addresses Pursuant to Resolution 102 (Minneapolis 1998)." Plenipotentiary Conference, Marrakesh. http://www.itu.int/osg/spu/mina/2002/inf4E.html.

Janisch, H. N. 1989. "The Canada-U.S. Free Trade Agreement: Impact on Telecommunications." *Telecommunications Policy* 13 (2): 99–103.

Jeon, Youngsuk. 1998. "A Study on Conflict between National and International Forces in Telecommunications Policymaking." Unpublished master's thesis, City University, London.

Johnson, Bobbie. 2005. "Rice Pressured EU over Internet Control." *Guardian*, 6 December.

Johnson, Chalmers. 2000. *Blowback: The Costs and Consequences of American Empire*. Reissued 2004. New York: Holt.

Johnson, Craig. 1991. "Is There Life Still in the Uruguay Round?" *Transnational Data and Communications Report* 14 (2): 6–8.

Joshi, Dinshaw F. D. 1987. "Availability of Financing for Telecommunications in Developing Countries." In *Small Earth Station Symposium and Exposition*, ed. Intelsat, 68–72. Washington, D.C.: Intelsat.

Kapur, Devesh. 2002. "The Changing Anatomy of Governance of the World Bank." In Pincus and Winters, 54–75.

Kazuka, Charles. 1989. "The International Telecommunication Union and Its Development Assistance Programme: A Perspective." Unpublished master's thesis, City University, London.

Kelly, Tim. 1994. "Global Regulator or Policy Forum." *Intermedia* 22 (4): 39.

———. 1997. "Ten Propositions for Accounting Rate Reform." Paper presented to ITU Asia Telecom, 13 June.

———. 1999. "Cost-Based and Demand-Based Tariffs." ALTTC Seminar, Ghaziabad, 20–22 July. http://www.itu.int/ITU-D/ict/ 1999/India/TK%20 tariffs%20Jul99.pdf.

Keohane, Robert O., and Helen V. Milner, eds. 1997. *Internationalization and Domestic Politics*. Cambridge: Cambridge University Press.

Keohane, Robert O., and Joseph S. Nye. 1977. *Power and Interdependence*. Boston: Little.

Kerver, Tom. 1989. "The Common Good." *Satellite Communications* 7 (4): 29–30.

Kildow, Judith T. 1973. *INTELSAT: Policy-Maker's Dilemma*. Lexington, Mass.: Lexington.

Kinn, Robert A. 1985. "United States Participation in the International Telecommunication Union: A Series of Interviews." *Fletcher Forum* 9 (Winter): 37–68.

Kinsley, Michael E. 1976. *Outer Space and Inner Sanctums: Government, Business and Satellite Communications.* New York: Wiley.

Kleinwächter, Wolfgang. 2003. "From Self-Governance to Public-Private Partnership: The Changing Role of Governments in the Management of the Internet's Core Resources." *Los Angeles Law Review* 36 (3): 1103–26.

———. 2004. "Beyond ICANN vs. ITU? How WSIS Tries to Enter the New Territory of Internet Governance." *Gazette: International Journal for Communication Studies* 66 (3–4): 233–51.

Klutznick, Philip M. 1988. "A Boost for the World Bank: Congress Shouldn't Hesitate." *Washington Post,* 29 September.

Krasner, Stephen D., ed. 1983a. *International Regimes.* Ithaca, N.Y.: Cornell University Press.

———. 1983b. "Structural Causes and Regime Consequences: Regimes as Intervening Variables." In Krasner, 1–22.

———. 1985. *Structural Conflict. The Third World against Global Liberalism.* Berkeley: University of California Press.

Krueger, Anne O., ed. 1998. *The WTO as an International Organisation.* Chicago: University of Chicago Press.

———. 1998. "An Agenda for the WTO." In Krueger *WTO as an International Organisation,* 401–10.

Lancaster, Carol. 1993. "Governance and Development, 1993: The Views from Washington." *IDS Bulletin* 24 (1): 9–15.

Lang, Tim, and Colin Hines. 1993. *The New Protectionism: Protecting the Future against Free Trade.* New York: New Press.

Lee, Kelley. 1996. *Global Telecommunications Regulation.* London: Pinter.

Leeson, Kenneth W., and C. Randall Jacobson. 1983. "Trade in Telecommunication Equipment and Services." In *Issues in International Telecommunications Policy: A Sourcebook,* ed. Jane H. Yurow, 180–88. Washington, D.C.: George Washington University.

Littlechild, Stephen C. 1983. *Regulation of British Telecommunications' Profitability.* London: Department of Industry.

Livingston, Steven. 1992. "The Politics of International Agenda Setting: Reagan and North-South Relations." *International Studies Quarterly* 36: 313–30.

Lloyd, Andrew. 1983. "Europe Pitches for African Communications Satellite." *New Scientist,* 18 June.

Loundy, David. 1997. "Obscure Experiment of Naming Domains." *CDLB Technology Law Column.* http://www.loundy.com/CDLB/IAHC-MoU.html.

Lukes, Steven. 2004. *Power: A Radical Analysis.* Harmondsworth, U.K.: Palgrave.

MacLean, D. J. 1995. "A New Departure for the ITU: An Inside View of the Kyoto Plenipotentiary Conference." *Telecommunications Policy* 19 (3): 177–90.

Mahoney, Eileen. 1993. "The Utilization of International Communications Organizations 1978–1992." In *Beyond National Sovereignty: International Communication in the 1990s,* eds. Kaarle Nordenstreng and Herbert I. Schiller, 314–42. Norwood, N.J.: Ablex.

Majtenyi, Cathy, and Michele Fleet. 1996. "Wiring Africa." *New Internationalist.* http://www.newint.org/issue286/wiring htm.

Mance, Oswald. 1943. *International Telecommunications.* London: Oxford University Press.

———. 1946. *Frontiers, Peace Treaties and International Organisation.* London: Oxford University Press.

Marsh, David. 1984. "Luxembourg Satellite TV Scheme Opposed as Anti-European." *Financial Times,* 31 May.

Mathew, Bobjoseph. 2003. *The WTO Agreements on Telecommunications.* Bern: Lang.

McCarthy, Kieren. 2006. "United States Cedes Control of the Internet—But What Now?" 27 July. http://www.theregister.co.uk/2006/07/27/ntia_icann_meeting.

McDowell, Stephen. 1997. *Globalization, Liberalization and Policy Change: A Political Economy Of India's Communications Sector.* Basingstoke, U.K.: Macmillan.

McKendrick, George G. 1987. "The INTUG View on the EEC Green Paper." *Telecommunications Policy* 11 (4): 325–29.

———. 2000. "INTUG: The Intug Story." http://www.intug.net/background/george_story.html.

McRae, Hamish, and Alex Brummer. 1985. "U.S. Sees a Role for the World Bank." *Guardian,* 9 October.

McWhinney, Edward, ed. *The International Law of Communications.* Leyden: Sijthoff.

Melody, William. 2000. "The Rise and Decline of the International Telecommunications Regime." In *Regulating the Global Information Society,* ed. Christopher Marsden, 124–77. London: Routledge.

Mendler, Camille, and Kenneth Hart. 1996. "Net Funding under Scrutiny." *Communications Week International,* 15 July: 3.

Merritt, Giles, and Paul Cheesewright. 1982. "Gatt Formula Depends on Last-Minute Chemistry." *Financial Times,* 24 November.

Mestmäcker, Ernst-Joachim. 1985. "The Impact of Deregulation on Trade in Services Agreements and on Telecommunications Monopolies in Europe." *The Washington Round: World Telecommunication Forum, 18–19 April,* 307–23. Geneva: ITU.

Meyers, Tedson J. 1992. "Seeking Mechanisms for Solving International Telecom Disputes." *Transnational Data and Communications Report* 15 (5): 9–11.

Mikesell, Raymond. 1972. "The Emergence of the World Bank as a Development Institution." In Acheson et al., 70–87.

Mirus, Rolf. 1989. "Canada-U.S. Free Trade Agreement." http://www.kanada-studien.de/Zeitschrift/zks18/Mirus.pdf.

Mistry, Percy. 1989. *The Present Role of the World Bank in Africa.* London: IFAA.

Molano, Walter T. 1997. *The Logic of Privatization: The Case of Telecommunications in the Southern Cone of Africa*. Westport, Conn.: Greenwood.

Moloney, David. 1996. "Developing Nations Dissent over Satellite Plans." *Communications Week International*, 4 November: 1, 39.

———. 1998a. "ITU Backs Up the Words with Action." *Communications Week International*, 6 April. http:// www.totaltele.com.

———. 1998b. "ITU Braced for Row over Standards." *Communications Week International*, 18 May: 1, 27.

Mondale, Leo. 1994. "Satellite Systems: A Challenge to Communication Needs in Rural and Remote Areas." Paper presented at ITU Africa Telecom '94, Cairo, Egypt. Geneva: ITU.

Montagnon, Peter. 1989. "World Bank Urged to Boost Backing for Privatisation." *Financial Times*, 16 May.

Moran, Michael. 2003. *The British Regulatory State*. Oxford: Oxford University Press.

Morgenthau, H. J. 1946. *Politics among Nations*. New York: Knopf.

Mosco, Vincent. 1990. "Toward a Theory of the State and Telecommunications Policy." In *Current Issues in International Communication*, eds. L. J. Martin and Ray Eldon Hiebert, 49–61. London: Longman.

Mosley, Paul, Jane Harrigan, and John Toye. 1991. *Aid and Power: The World Bank and Policy-Based Lending, Vol. I*. London: Routledge.

Motohiro, T., and A. Thierer. 2003. "Is America Exporting Misguided Telecommunications Policy? The U.S.-Japan Telecom Trade Negotiations and Beyond." Cato Institute Briefing Paper No. 79. http://www.cato.org.

Mullins, Steve. 1995. "Asia, Africa Give Boost to GSM." *Communications Week International*, 8 September: 55–57.

Murphy, Thomas P. 1971. "Technology and Political Change: The Public Interest Impact of Comsat." *Review of Politics* 33 (4): 10–20.

Musolf, Lloyd. 1968. *Communications Satellites in Political Orbit*. San Francisco: Chandler.

Naftel, Mark, and Lawrence Spivak. 2000. *The Telecoms Trade War: The United States, the European Union and the World Trade Organisation*. Oxford: Hart.

Naraine, Mahindra. 1985. "Direct Broadcasting Satellites: New Technologies and Traditional Concepts." Paper presented to the British International Studies Association, Bristol, 7 December.

———. 1986. "Competition in International Satellite Communications—An Historical Analysis." In *Telecommunications: National Policies in an International Context*, ed. Nicholas Garnham, 257–302. London: Polytechnic of Central London.

Näslund, Ruben. 1983. "ITU Conference in Nairobi." *Telecommunications Policy* 7 (2): 100–110.

Nelson, Robert A. 1998. "Iridium: From Concept to Reality." http://www.aticourses .com/news/iridium.htm.

Neu, Werner, and Thomas Schnöring. 1989. "The Telecommunications Equipment Industry: Recent Changes in Its International Trade Pattern." *Telecommunications Policy* 13 (3): 25–37.

Newman, Eugene. 1993. "The Challenge of International Value-Added Network Services." *Telecommunications Policy* 17 (5): 370–78,

Newman, Mark. 1992. "Privatisation Programmes: Momentum Remains Strong." *Financial Times,* 15 October.

NGO Working Group on the World Bank. 1989. *Position Paper of the NGO Working Group on the World Bank.* Geneva: ICVA.

Niskanen, William A. 1987. "Stumbling toward a U.S.-Canada Free Trade Agreement." Cato Institute. http://www.cato.org.

Noam, Eli. 1987. "The Public Telecommunications Network: A Concept in Transition." *Journal of Communication* 37 (1): 30–48.

———. 1989. *International Telecommunications in Transition: The Breakdown of Anomaly.* New York: Columbia University, Center for Telecommunications and Information Studies.

Noam, Eli M., and Gerald Pogorel, eds. *The Dynamics of Telecommunications Policy in Europe and the United States.* Norwood, N.J.: Ablex.

Nogués, Julio J. 1998. "The Linkages of the World Bank with the GATT/WTO." In Krueger, *WTO as an International Organisation,* 82–95.

Noll, Alfons. 1985. "The Institutional Framework of the ITU and its Various Approaches with Regard to International Telecommunication Law and Treaty Conferences in ITU." *The Washington Round: World Telecommunication Forum,* 18–19 April, 19–41. Geneva: ITU.

Nora, Simon, and Alain Minc. 1980. *The Computerization of Society.* Cambridge, Mass.: MIT Press.

Nordenstreng, K. 1984. *The Mass Media Declaration of UNESCO.* Norwood, N.J.: Ablex.

Nulty, Timothy E. 1989. "Emerging Issues in World Telecommunications." In Wellenius, P. Stern, Nulty, and R. Stern, 7–18.

Nye, Joseph S. 1969. "Unctad: Poor Nations' Pressure Group." In *The Anatomy of Influence,* eds. Robert W. Cox and Harold K. Jacobson, 334–70. London: Yale University Press.

———. 2004. *Power in the Global Information Age.* London: Routledge.

Nye, Sheridan. 1997. "ITU Battles FCC for Moral High Ground on Accounting Rates." 5 December. http://www.totaltele.com.

Odell, John, and Barry Eichengreen. 1998. "The United States, the ITO, and the WTO: Exit Options, Agent Slack, and Presidential Leadership." In Krueger, *WTO as an International Organisation,* 181–212.

OECD (Organisation for Economic Co-operation and Development). 1994. *International Telecommunication Tariffs.* Paris: OECD.

Oettinger, Anthony G. 1988. *The Formula Is Everything: Costing and Pricing in the Telecommunications Industry.* Cambridge, Mass.: Center on Information Resources Policy, Harvard University.

Oftel. 2000. *International Controls in PTO Licences: Consultation Document.* London: Oftel.

————. 2002. *The Use of Long Run Incremental Cost (LRIC) as a Costing Methodology in Regulation.* http://www.ofcom.org.uk/static/archive/Oftel.publication/mobile/ctm_2002/lric120202.pdf.

Ohmae, Kenichi. 1995. *The End of the Nation State.* New York: Free Press.

Olson, M. 1965. *The Logic of Collective Action: Public Goods and the Theory of Groups.* Cambridge, Mass.: Harvard University Press, 1978.

Olufs, Dick. 1999. *The Making of Telecommunications Policy.* Boulder, Colo.: Reiner.

O'Rorke, Richard J. 1985. *Estimation of the Population of Large International Private Lease Users.* U.S. Department of Commerce, National Telecommunications and Information Administration. Washington, D.C.: GPO.

Ó Siochrú, Seán. 1995. "International Telecommunication Union and Non-Governmental Organisations: The Case for Mutual Cooperation." Geneva: ITU. http://www.comunic.org/itu_ngo/mutu.

Oslund, Jack. 1977. "'Open Shores' to 'Open Skies': Sources and Directions of U.S. Satellite Policy Environment." In Pelton and Snow, 143–99.

OUT-LAW News. 2006. "U.S. to Liberate ICANN by 2009." 3 October. http://www.out-law.com/default.aspx?page=7355.

Oxfam. 1993. *Africa: Make or Break; Action for Recovery.* Oxford: Oxfam.

Palast, Gregory. 1998. "A Marxist Threat to Cola Sales? Pepsi Demands a U.S. Coup: Goodbye Allende, Hello Pinochet." *Observer*, 8 November.

Pastore, J. O. 1964. *The Story of Communications.* New York: Macfadden-Bartell.

Payer, Cheryl. 1982. *The World Bank: A Critical Analysis.* London: Monthly Review.

Peet, Richard. 2003. *Unholy Trinity: The IMF, World Bank and WTO.* Malaysia: SIRD.

Pelton, Joseph N. 1977. "Key Problems in Satellite Communications: Proliferation, Competition and Planning in an Uncertain Environment." In Pelton and Snow, 93–142.

————. 1984a. "Intelsat, Communications Development and World Communications Year." *Telematics and Informatics* 1 (1): 75–85.

————. 1984b. "Intelsat and Initiatives for Third World Development." In *World Communications: A Handbook,* eds. George Gerbner and Marsha Siefert, 411–17. New York: Longman.

Pelton, Joseph N., and Marcellus S. Snow, eds. 1977. *Economic and Policy Problems in Satellite Communications.* New York: Praeger.

Petrazzini, Ben. 1996. *Global Telecom Talks: A Trillion Dollar Deal.* Washington, D.C.: Institute for International Economics.

Pfeffermann, Guy. 1992. "Facilitating Foreign Investment." *Finance and Development* March: 46–49.

Pincus, Jonathan R., and Jeffrey A. Winters, eds. *Reinventing the World Bank.* Ithaca, N.Y.: Cornell University Press.

Pipe, G. Russell. 1989. "The ITU Plenipot: Renewal without Reform." *Transnational Data and Communications Report* 12 (7): 5–15.

———. 1993. "Trade of Telecommunications Services: Implications of a GATT Uruguay Round Agreement for ITU and Member States." Geneva: ITU.

Pirie, Madsen. 1988. *Privatization, Theory, Practice and Choice.* London: Wildwood.

Powell, Michael. 2004. "The Age of Personal Communications: Power to the People." Speech. National Press Club, 14 January. http://hraunfoss.fcc.gov/edocs_public/attachmatch/DOC-242885A1.doc.

Priddle, Robert. 1989. "The United Kingdom." *Conference on Telecommunications and the Melbourne Meetings,* 31 March. London: IBC.

Prowse, Michael. 1991. "'Tough Manager' Stamps His Authority on World Bank." *Financial Times,* 19 September.

Purton, Peter. 1987. "Theodor Irmer Copes with Change at the CCITT." *Telephony* 28 September: 152–54.

Quello, James H. 2001. *My Wars: Surviving WWII & the FCC.* Arlington, Va.: Alexis de Tocqueville Institution.

Raghavan, Chakravarthi. 1990. *Recolonization, GATT, the Uruguay Round and the Third World.* London: Zed.

———. 1993a. "Last-Ditch Effort to Meet Mid-January Deadline." 29 January. http://www.sunsonline.org.

———. 1993b. "Burning the Midnight Oil for a Cancelled Examination." 29 January. http://www.sunsonline.org.

———. 1993c. "No Official Nominations for GATT DG, but . . . ." 22 April. http://www.sunsonline.org.

Raveendran, Laina. 1989. "WATTC-88." In *Reforming the Global Network,* ed. IIC, 25–27. London: IIC.

Renaud, Jean-Luc. 1985. "U.S. Government Assistance to Associated Press's Worldwide Expansion." *Journalism Quarterly* 62: 10–16.

———. 1987. "The ITU and Development Assistance." *Telecommunications Policy* 11 (4): 179–91.

Rhodes, R. A. W. 1996. "The New Governance: Governing without Government." *Political Studies* XLIV: 652–67.

Rhodes, R. A. W., and D. Marsh, eds. 1992. *Policy Networks in British Governments.* Oxford: Clarendon.

Rich, Brian. 2002. "The World Bank under James Wolfensohn." In Pincus and Winters, 26–53.

Riddell, Peter. 1989. "U.S. Trade Armoury Gets A 'Crowbar.'" *Financial Times,* 2 February.

Rischard, J. F. 1995. "World Bank Announces InfoDev Fund at GIIC Forum." *Iways: Digest of the Global Information Infrastructure Commission* March/April: 47–52.

Ritzen, Josef. 2005. *A Chance for the World Bank.* London: Anthem.

Robinson, Peter. 1985. "Telecommunications, Trade and TDF," *Telecommunications Policy* 9 (4): 311–18.

Roseman, Daniel. 2005. "The WTO and Telecommunications Services in China: Three Years On." *Info* 7 (2): 25–48.

Rowen, Hubert. 1987. "Clausen Asks More Private Sector Aid." *Washington Post*, 23 December.

———. 1991. "U.S. Eases Campaign to Alter World Bank." *Washington Post*, 7 June.

Rutkowski, A. M. 1984. "The ITU and the U.S.: Partners or Rivals?" In Sterling, 28–39.

———. 1986a. "The New International Telecommunication Regulations." *Telecommunications* 20 (1): 66–72.

———. 1986b. "Regulation for Integrated Services Networks: WATTC-88." *Intermedia* 14 (3): 10–19.

———. 1998. "A Lost Cause for National Policymakers." *Communications Week International* (14 December): 8.

———. 1999. "Telecom Rules Cannot Apply to the NET" *Communications Week International*, 4 October: 10.

Ryrie, William. 1991. "Free-Market Preaching in a World with No Marketplace." *Washington Post*, 28 July.

Sampson, Anthony. 1973. *The Sovereign State: The Secret History of ITT.* London: Hodder.

Saunders, Robert J. 1982. "Telecommunications in Developing Countries: Constraints on Development." In *Communications and Economic Development*, eds. M. Jussawalla and D. M. Lampton, 190–209. London: Pergamon.

Saunders, Robert J., Jeremy J. Warford, and Björn Wellenius. 1983. *Telecommunications and Economic Development.* Washington, D.C.: World Bank.

Sauvant, Karl P. 1983. "Transborder Data Flows and Developing Countries." *International Organisation* 37 (2): 359–71.

Schaffer, Jon. 1998. "U.S. Willing to Show Flexibility on Global Telecom Issues." USIS Washington File. http://canberra.usembassy.gov/hyper/WF980313/epf512.htm.

Schenker, Jennifer, and Karen Lynch. 1991. "Resale a Means to Drive Down Prices." *Communications Week International*, 16 December.

Schiller, Dan. 1982. "Business Users and the Telecommunications Network." *Journal of Communication* 32 (Autumn): 84–96.

Schiller, Herbert I. 1969. *Mass Communications and American Empire.* 2nd ed. Boulder, Colo.: Westview.

———. 1976. *Communication and Cultural Domination.* New York: Sharpe.

———. 1979. "Genesis of the Free Flow of Information Principles." In *Communication and Class Struggle: Capitalism, Imperialism*, vol. 1, ed. Armand Mattelart and Seth Siegelaub, 345–53. New York: Intl. Mass Media.

Schniad, Sid. 1992a. "GATT, the Canada-U.S. Free Trade Agreement, NAFTA and the Corporate Game Plan." Paper, 3rd Coquio de Xalapa, *Restructuracion Productiva y Reorganizacion Social,* 7–10 October.

———. 1992b. "NAFTA Undermines Canadian Control over Communications." Paper for Telecommunications Workers Union, Burnaby, B.C., September.

Schoonmaker, Sara. 2002. *High-Tech Trade Wars: U.S.-Brazilian Conflicts in the Global Economy.* Pittsburgh: University of Pittsburgh Press.

Scott-Joynt, Jeremy. 1997. "C&W Set to Challenge FCC in Court over Accounting Rate Cuts." 17 August. http://totaltele.com.

Sebesta, Lorenza. n.d. [circa 1995]. "Chapter 11: U.S.-European Relations and the Decision to Build Ariane, the European Launch Vehicle." http://history.nasa.gov/SP-4217/ch11.htm.

Sherman, Laura B. 1998. "'Wildly Enthusiastic' about the First Multilateral Agreement on Trade in Telecommunications Services." *Federal Communications Law Journal* 51 (1) (December). http://www.law.indiana.edu/fclj/pubs/v51no1.html.

Shetty, Vineeta. 1998a. "ITU in Knots over Accounting Rates." 2 November. http://www.totaltele.com.

———. 1998b. "ITU Conference: Show Me the Money." 1 December. www.totaltele.com.

Shurmer, Mark, and Paul A. David. 1996. "Formal Standards-Setting for Global Telecommunication and Information Services: Towards Institutional Renovation or Collapse?" London: ESRC.

Singh, J. P. 2002. "Negotiating Regime Change: The Weak, the Strong and the WTO Telecom Accord." In *Information Technologies and Global Politics,* eds. James N. Rosenau and J. P. Singh, 239–74. Albany, N.Y.: SUNY Press.

Smith, Delbert D. 1976. *Communication via Satellite: A Vision in Retrospect.* Boston: Sijthoff.

Smith, Michael B. [Deputy U.S. Trade Representative]. 1982. "USIA Interview with Ambassador Smith, U.S. Trade Representative in Geneva," U.S. Information Service press release, 15 November.

Smith, Peter L., and Gregory Staple. 1994. *Telecommunications Sector Reform in Asia: Towards a New Pragmatism.* Washington, D.C.: World Bank.

Snoddy, Raymond. 1985. "Luxembourg TV Satellite Scheme Excludes Coronet." *Financial Times,* 2 March.

Snow, Marcellus S. 1976. *International Commercial Satellite Communications.* New York: Praeger.

———. 1986. "Communications Policy in Seven Developed Countries: Introduction, Background and Conclusions." In *Telecommunications Regulation and Deregulation in Industrialized Democracies,* ed. Marcellus S. Snow, 3–9. Amsterdam: North Holland.

Sobel, Robert. 1982. *ITT: The Management of Opportunity.* London: Sidgwick.

Solomon, Jonathan. 1984. "The Future Role of International Telecommunications Institutions." *Telecommunications Policy* 8 (3): 213–22.

Spero, Joan Edelman. 1982. "Information: The Policy Void." *Foreign Policy* 48 (Fall): 139–56.

———. 1999. "How the FCC is Overstepping Its Mark." *Communications Week International,* 15 February: 11.

Spivak, Lawrence J. 1999. "How the FCC Is Overstepping Its Mark" *Communications Week International,* 15 February.

Stanley, Kenneth B. 1977. "Economic Issues in International Telecommunications: A Public Policy Dilemma." In Pelton and Snow, 62–88.

———. 1988. *The Balance of Payments Deficit in International Telecommunications Services.* Washington, D.C.: Common Carrier Bureau, FCC.

Sterling, C. H., ed. *International Telecommunications and Information Policy.* Washington, D.C.: Ctr. for Telecom. Studies, George Washington Univ.

Stern, Peter A., and Tim Kelly. 1997. "Liberalization and Reform of International Telecommunication Settlement Arrangements." Latin American and Caribbean Telecommunication Finance and Trade Colloquium, 14–16 July.

Stern, Richard D. 1986. "The World Bank's Role in Fostering Telecommunications Development." In *Telecommunications for Development: An International Forum,* ed. Intelsat, 166–72. Washington, D.C.: Intelsat.

———. 1989. "Alternatives for the Future." In Wellenius, P. Stern, Nulty, and R. Stern, 125–29.

Stigler, George. 1975. *The Citizen and the State: Essays on Regulation.* Chicago: University of Chicago Press.

Stiglitz, Joseph E. 2002. *Globalization and its Discontents.* New York: Norton.

Strange, Susan. 1983. "Cave hic dragones: A Critique of Regime Analysis." In Krasner, 337–55.

Summers, Lawrence. 1992. "The Challenge of Development: Some Lessons of History for Sub-Saharan Africa." *Finance and Development,* March: 6–9.

Sussman, Gerald. 1991. "Telecommunications for Transnational Integration: The World Bank in the Phillipines." In *Transnational Communications: Wiring the Third World,* eds. Gerald Sussman and J. Lent, 28–41. London: Sage.

Tarjanne, Pekka. 1992a. "Telecom: Bridge to the 21st Century." *Transnational Data and Communications Report* 15 (3): 42–45.

———. 1992b. "The ITU Responds to New Concepts for Public Policy in the Global Information Society." *Intermedia* 20 (6): 12–14.

———. 1997. "The WTO Basic Telecommunications Agreement: An ITU Viewpoint." Conference on Negotiating International Trade Issues, 28 November. Geneva: WTO. http://www.itu.int/itudoc/osg/ptspeech/chron/1997/42248.html.

Taylor, Leslie A. 1989. "WARCs, WATTCs & Plenipots." *Via Satellite* (May): 5–6.

Third World Economics. 1990. "African Ministers Concerned over Trend in Trade Talks." 1–15 May: 4.

Thompson, Geoffrey. 1981. "Challenging CEPT Agreements." *Telephony,* 26 January.

Thomson, Robert. 1990. "Japan Responds with Calm to Gatt Crisis." *Financial Times,* 18 December.

Thuswaldner, Andreas. 2000. "The International Revenue Settlement Debate." *Telecommunications Policy* 24: 31–50.

Tiger, Michael. 1992. "NAFTA's New Code Book for the World's Largest Market." *Intermedia* 20 (6): 17–18.

Tran, Mark. 1989. "'Super 301' Puts U.S. Trade Rivals on Fear Footing." *Financial Times,* 4 May.

———. 1994. "Bentsen in Searing World Bank Attack." *Guardian,* 27 April.

Transnational Data and Communications Report. 1989. "Telecom Central to Services Negotiations." 12 (10): 5.

———. 1990."International Telecom Charging Reforms Ripe." 13 (6): 5–6.

———. 1991a. "Telecom Important to Restarted Services Negotiations." 14 (2): 5–6.

———. 1991b. "Accomplishments Expected from GATT Trade in Services." 14 (6): 3–4.

———. 1992. "U.S. Formally Exempts Basic Telecom from Uruguay Round." 15 (3): 5.

———. 1993a. "Basic Telecom Coverage under GATS Proposed." 16 (1): 5–6.

———. 1993b. "All Telecom Market Openings Favour US Interests." 16 (5): 5–7.

———. 1994. "Pricing Dropped from Telecommunications Annex." 17 (1): 7.

Tricks, Henry. 1998. "AT&T Escalates Telmex Dispute." *Financial Times,* 9 February.

Tritt, Robert. 1992. "GATS, Son of GATT: A New Rule Book for Cross-Border Competition." *Intermedia* 20 (6): 15–16.

Tyler, Christian. 1984. "Why Gatt Lacks International Clout." *Financial Times,* 13 June.

Tyler, Christian, and William Dullforce. 1986. "When Failure is Unaffordable." *Financial Times,* 15 September.

Tyler, Michael. 1995. *Global Mobile Personal Communications Systems.* ITU Regulatory Colloquium 3, November 1994. Geneva: ITU.

———. 1998. *Transforming Economic Relationships in International Telecommunications.* ITU Regulatory Colloquium No. 7. Geneva: ITU.

Ungerer, Herbert. 1989. "WATTC-88: The European Community." Conference on Telecommunications and the Melbourne Meetings, 31 March. London: IBC.

United Kingdom, Department of Trade and Industry. 1993. "Multilateral Development Agencies—UK Procurement." World Aid Section. October.

United Nations Centre on Transnational Corporations. 1983. *Transborder Data Flows and Brazil.* New York: UN.

Urey, Gwen. 1995. "Infrastructure for Global Financial Integration: The Role of the World Bank." In *Telecommunications Politics: Ownership and Control of the Information Highway in Developing Countries,* eds. Bella Mody, Johannes M. Bauer, and Joseph D. Straubhaar, 113–34. Mahwah, N.J.: Erlbaum.

U.S. Congress. House of Representatives. 1963. *Military Communications Satellite Program.* Hearing before a Subcommittee of the Committee on Government Operations. 88th Cong., 1st sess., 23 April. Washington, D.C.: GPO.

———. 1965. *Satellite Communications (Military-Civil Roles and Relationships).* Second Report by the Committee on Government Operations. 89th Cong., 1st Sess. Washington, D.C.: GPO.

———. 1981. *International Broadcasting.* Hearings before a Subcommittee of the Committee on Government Operations. 97th Cong., 1st Sess., September 16. Washington, D.C.: GPO.

———. 1982. *International Telecommunications and Information Policy.* Hearings before Subcommittee of the Committee on Government Operations. 97th Cong., 1st and 2nd sess., 2 December 1981 and 29 April 1982. Washington, D.C.: GPO.

———. 1983. *International Trade Issues in Telecommunications and Related Industries.* Hearing before the Subcommittee on Telecommunications, Consumer Protection, and Finance of the Committee on Energy and Commerce. 98th Cong., 1st Sess., 23 March. Washington, D.C.: GPO.

U.S. Congress. Senate. 1950. *Annex to International Telecommunication Convention—Telegraph Regulations (Paris Revision 1949) and Final Protocol. Submission for Ratification 10 April 1950.* 81st Cong., 2nd Sess.,Washington, D.C.: GPO.

———. 1983. Committee on Foreign Relations. Report prepared by Library of Congress. *International Telecommunications and Information Policy: Selected Issues for the 1980s.* Washington, D.C.: GPO.

U.S. Department of Commerce. NTIA. 1990. *U.S. Telecommunications in a Global Economy.* Washington, D.C.: GPO.

U.S. Department of State. Bureau of International Communications and Information Policy. 1989. *Report of the Delegation of the United States to the Plenipotentiary Conference of the International Telecommunication Union, Nice, France, May 23–June 30 1989.* Washington, D.C.: GPO.

USITUA (United States ITU Association) Newsletter. 2000. "The Point" No. 1. http://www.usitua.org/point-home.htm.

———. 2003. "The Point" No. 8. http://www.usitua.org/point-home.htm.

U.S. White House. 1998. *The Framework for Global Electronic Commerce: Read the Framework.* http://clinton4.nara.gov/WH/New/Commerce/read.html.

Utsumi, Yoshio. 2000. "Moving beyond International Accounting Rates." *Telecommunications Policy* 24 (1): 5–8.

Valentine, W. R. 1981. "Policy Implications of the World Administrative Radio Conference of 1979: An Historical Perspective and Political Analysis." Unpublished master's thesis, Michigan State University, East Lansing.

Vines, David. 1998. "The WTO in Relation to the Fund and the Bank: Competencies, Agendas and Linkages." In Krueger, *WTO as an International Organisation*, 59–82.

Vogler, John. 2000. *The Global Commons.* London: Wiley.

Wallsten, Scott. 2002. "Does Sequencing Matter? Regulation and Privatizations in Telecommunications Reform." Policy Research Working Paper. Series 2817. Washington, D.C.: World Bank.

Watkins, Kevin. 1992. *Fixing the Rules: North-South Issues in International Trade and the Gatt Uruquay Round.* London: Catholic Inst. for Intl. Relations.

Weinberg, Philip. 1997. "Intelsat's Future Structure." Unpublished master's thesis, City University, London.

Wellenius, Björn. 1984a. "Economic Analysis of Telecommunications Investment In Developing Countries." *World Telecommunications Forum* 3 (3): 1–5. Geneva: ITU.

———. 1984b. "Telecommunications in Developing Countries." *Finance and Development* (September): 33–35.

———. 1986. "Financing Telecommunications in the Developing World—Issues and Opportunities." *Africa Telecom '86,* 1–5. Geneva: ITU.

———. 2002. *Closing the Gap in Access to Rural Communications.* World Bank Discussion Paper 430. Washington, D.C.: World Bank.

Wellenius, Björn, Juan Galarza, and Boutheina Guermazi. 2005. "Telecommunications and the WTO. The Case of Mexico." Washington, D.C.: World Bank Global Information and Communications Technologies. http://www.worldbank.org.

Wellenius, Björn, and others. 1993. *Telecommunications: World Bank Experience and Strategy.* Washington, D.C.: World Bank.

Wellenius, Björn, Peter A. Stern, T. E. Nulty, and Richard Stern. 1989. *Restructuring and Managing the Telecommunications Sector.* Washington, D.C.: World Bank.

Whalen, David J. n.d. "Billion Dollar Technology: A Short Historical Overview of the Origins of Communications Satellite Technology 1945–65." http://History.nasa.gov/SP-4217/ch9.htm.

———. 2002. *The Origins of Satellite Communications 1945–1965.* Washington: Smithsonian.

Wigglesworth, W. R. B. 1997. "The Role of Information in Telecom Regulation." In *Telecom Reform: Principles, Policies and Regulatory Practices,* ed. William H. Melody, 295–315. Denmark: Technical University of Denmark.

Williams, David, and Tom Young. 1994. "Governance, the World Bank and Liberal Theory." *Political Studies* 42 (1): 84–100.

Williams, Frances. 1994. "Gatt's Successor to be Given Real Clout Move." *Financial Times,* 7 April.

———. 1996. "Global Deal on Telecoms Threatened by U.S. Move." *Financial Times,* 23 April.

———. 1997. "Trade Chief Appeals for Telecom Deal." *Financial Times,* 11 February.

Winsbury, Rex. 1994. "Who Will Pay for the Global Village? Funding the Buenos Aires Declaration." *Intermedia* 22 (3): 23–31.

Winseck, Dwayne. 1995. "Power Shift? Towards a Political Economy of Canadian Telecommunications and Regulation." *Canadian Journal of Communications* 20 (1). http://www.cjc-online.ca/viewarticle.php?id=269&layout=html.

———. 2002. "The WTO, Emerging Policy Regimes and the Political Economy of

Transnational Communications." In *Global Media Policy in the New Millenium*, ed. Marc Raboy, 19–38. Luton, U.K.: University of Luton Press.

Winston, Brian. 1998. *Media, Technology and Society*. London: Routledge.

Woodrow, Brian R., and Pierre Sauvé. 1994. "The European Community and the Uruguay Round Services Trade Negotiations." In *Telecommunications in Transition*, eds. Charles Steinfeld, Johannes M. Bauer, and Laurence Caby, 97–117. London: Sage.

World Bank. 1981. *Accelerated Development in Sub-Saharan Africa*. Washington, D.C.: World Bank.

———. 1989. *Sub-Saharan Africa: From Crisis to Sustainable Growth*. Washington, D.C.: World Bank.

———. 1993a. *Argentina's Privatization Program: Experience, Issues and Lessons*. Washington, D.C.: World Bank.

———. 1993b. *The East Asian Miracle: Economic Growth and Public Policy*. Oxford: Oxford University Press.

———. 1994. *World Development Report: Infrastructure for Development*. Oxford: Oxford University Press.

———. 1997. *World Development Report: The State in a Changing World*. Oxford: Oxford University Press.

———. 2005. *Telecom Projects (1990–2005)*. Washington, D.C.: World Bank. http://web .worldbank.org/external/projects/main?pagePK=218616&piPK=217470.

Young, Oran R., and Gail Osherenko. 1995. "Testing Theories of Regime Formation." In *Regime Theory and International Relations*, ed. Volka Rittberger, 223–51. Oxford: Oxford University Press.

Young, Peter. 1988. *The Enterprise Imperative: Promoting Growth in Developing Countries*. London: Adam Smith.

Yurow, Jane H. 1983. *Issues in International Telecommunications Policy: Sourcebook*. Washington, D.C.: George Washington University.

Zacher, M., and B. Sutton. 1997. *Governing Global Networks: International Regimes for Transportation and Communications*. Cambridge: Cambridge University Press.

Zarkin, M. J. 2003. "Telecommunications Policy Learning: The Case of the FCC's Computer Inquiries." *Telecommunications Policy* 27: 283–99.

Zutshi, B. K. 1994. "GATS—Impact on Developing Countries and Telecom Services." *Transnational Data and Communication Report* 17 (4): 16–25.

JILL HILLS is the author of several books, including *The Struggle for Control of Global Communication, Deregulating Telecoms,* and *Information Technology and Industrial Policy.* Her articles have appeared in *Media Development, Polity, Review of International Political Economy, Review of International Studies, Telecommunications Journal, Telecommunications Policy,* and other journals.

THE HISTORY OF COMMUNICATION

The University of Illinois Press
is a founding member of the
Association of American University Presses.

———————————————————————

Composed in 10.5/13 Adobe Minion Pro
by Jim Proefrock
at the University of Illinois Press
Manufactured by Thomson-Shore, Inc.

University of Illinois Press
1325 South Oak Street
Champaign, IL 61820-6903
www.press.uillinois.edu